生命之水

艾雷岛威士忌品鉴指南：汉英对照

The Water Of Life : A Tasting Guide to Islay Whisky : Chinese-English Bilingual Version

［克罗］何沃德（Hrvoje Vojvodic） 编著

覃雨洋 编译

SPM 南方出版传媒 广东人民出版社

·广州·

图书在版编目（CIP）数据

生命之水：艾雷岛威士忌品鉴指南：汉英对照 / （克罗）何沃德 (Hrvoje Vojvodic) 编著；覃雨洋编译. — 广州：广东人民出版社，2021.1
ISBN 978-7-218-14018-6

Ⅰ. ①生… Ⅱ. ①何… ②覃… Ⅲ. ①威士忌酒－品鉴－指南－汉、英 Ⅳ. ①TS262.3-62

中国版本图书馆CIP数据核字(2019)第252780号

SHENG MING ZHI SHUI ： AILEIDAO WEISHIJI PINJIAN ZHINAN : HANYING DUIZHAO

生命之水：艾雷岛威士忌品鉴指南:汉英对照

[克罗] 何沃德 (Hrvoje Vojvodic) 编著　覃雨洋 编译　　

出 版 人：肖风华

选题策划：李　敏
责任编辑：李　敏
装帧设计：刘焕文
责任技编：吴彦斌　周星奎

出版发行：广州市海珠区新港西路204号2号楼（邮政编码：510300）
电　　话：（020）83798714（总编室）
传　　真：（020）83780199
网　　址：http://www.gdpph.com
印　　刷：广州市人杰彩印厂
开　　本：787 mm × 1092 mm　1/16
印　　张：20　　　　　　　字　　数：280千
版　　次：2021年1月第1版
印　　次：2021年1月第1次印刷
定　　价：268.00元

谨将本书献给想加入威士忌世界，或只是想尝尝这种佳酿的人们。

This book is for the people who want either to become part of whisky world or just taste a bit of it.

关于本书

我想把这本书做成一本对艾雷岛威士忌有一定认识，或希望接触到这类威士忌的人的指南。

在本书第一部分，你会读到关于艾雷岛的一些基本情况。地理、交通、有趣的地方等等。第二部分聚焦在艾雷岛漫长的历史上。第三部分会给你一些关于艾雷岛威士忌漫长历史的有趣信息。第四部分简短地介绍了艾雷岛威士忌的主要特点，同时介绍了现存的蒸馏厂名单。在第五部分也是本书的主要部分中，你会找到一份所有现存岛上的蒸馏厂以及旗下所有正在生产的威士忌的清单。你也可以在这里读到一些来自苏格兰各地的生产艾雷岛威士忌的公司。

每款威士忌都配有蒸馏厂的品鉴笔记，它们将帮助你理解它们之间的异同。

希望你能找到一些属于你的东西。

本书最后两个部分介绍了参考资料，并介绍了所有用知识和实际行动支持我编成此书的人们。

我希望你会享受其中，如同我享受写它的过程一样。

何沃德

About the book:

I want to make this book a guide to the people who already have some knowledge about the Islay type of whisky, as well as people who will get in touch with it.

In the first part of the book you will find some basic facts about Islay. Geography, transport, interesting locations, etc. The second part is dedicated to the long history of the island and the third part will give you some interesting information about the long history of Islay whisky. The fourth part will give you brief introduction of main characteristics of Islay whisky as well as list of today's distilleries. In the fifth part, the main one, you will find list of all existing distilleries on the island, with all whiskies which are currently in production. Also here you can see companies from all around the Scotland which are producing Islay whisky.

All the whiskies are provided with distilleries tasting notes which will help you understand the similarities and differences.

Hope you will find something for yourself.

The Last two parts of the book are reserved for references and to all the people who helped me to make this book. With their knowledge or by their support.

I wish you enjoy reading this book as much as I enjoyed making it for all of you.

Hrvoje Vojvodic

序言

记忆中最早接触威士忌是那杯"威水"，威士忌加水，这是威士忌爱好者的一种饮用方式，在粤港澳一带，"威水"还有另一个意思就是"威武"。

20世纪80年代，国内酒店的改革和国际酒店的建立，港澳地区起到很大的带动作用，当中很多的饮食专业术语，也常常用到粤语和港澳的行业用语来命名。我在那时候加入了一间国际五星级酒店，当时以香港酒店的管理文化为基础，也在那个时候我与威士忌结缘，从简单的威水，威士忌加冰开始认识威士忌。回首前尘，颇有感慨，正所谓杯酒人生，酒的味道可能模糊，但当中却充满了一页页的回忆。

夕阳斜落在酒店大堂的一个角落，这个是大堂的一个酒吧，叫"角落吧"。角落永远给你一种神秘感，酒吧开设在这角落，也给人带来几分好奇。在每天下午的5点半左右，记忆里总是坐着不少的外国人面孔，那时的外国人多数是改革开放之后，外国公司驻华的一些工作人员，他们把happy hour（傍晚酒聚）的生活习惯带来了本地。高高的吧凳可以半坐半站，吧台下边都有一个小钩，方便女士挂手袋。而吧台上边有一个铃铛，是给所谓的绅士们去"敲铃"的，此举是让敲铃人给所有现场的客人买单提供的信号，英式酒吧都有这个玩意，所以不能乱敲，就算小朋友敲了家长也要买单。这种英式的酒吧，与后来的威士忌吧基本上别无二致，后者只是多了一些"距离感"而已。

"毡威地伏冧"是对西式高度酒在命名上的一种简单概括，不过毡酒、伏特加、威士忌、冧酒、白兰地，这几种高度酒也可视作衡量饮用者对酒吧的认知程度，比如对鸡尾酒的了解，又或是酒与酒之间的配搭以及当中的风味、口感特点，乃至一些背景故事等，这些都是在酒吧里度过美好时光的回忆和谈资。酒作为一种媒介，令人的思维活跃，增加了人和人之间的交流艺术。而另一个重要的衡量方式就是，成熟的酒客喜欢喝威士忌，而且老酒鬼总是说，从啤酒转喝威士忌是一种成熟的表现。我觉得并无道理，有些酒鬼一辈子都不喝威士忌的，当然不少酒鬼说威士忌是世界上最香的酒。

我一直反对把白兰地笼统地命名为洋酒，但这种美丽的误会，到今天已经不能改变，小朋友对着白兰地也叫洋酒了。不过这也造就了威士忌的独立命名，和后来对酒品的细分，其中的分门别类比洋酒（白兰地）来得更加细致，大家在书里将会感受到作者的笔耕。

威士忌最堕落的时候莫过于威士忌加绿茶的那段日子，曾几何时，这种饮用方式有好几年在酒吧里大量的泛滥。而这种模式在各种商业手法的推动下，达到一定的市场效应，不过最后也寿终正寝。经过一段冷静期之后，慢慢地有人注意到"单一麦芽"的存在。国人的消费心理在各种内外因的种种推动下，使得威士忌再一次受到重视和关注，虽然还是小众的比例，但品牌的炒作和价格上涨，不难看到市场存在的价值。而且威士忌多数以一种绅士的形象传播，某种程度而言令本来在原产地是普罗大众的消费品变为高深莫测的奢侈品，致使人们开始从迷信、盲从，继而进入学习阶段。本书中以全新的视角介绍了艾雷岛威士忌的发展史、生产工艺和蒸馏厂历史，解读了岛上的九大蒸馏厂和公司，并分享了近100款经典品牌苏格兰威士忌的品鉴知识，让威士忌爱好者和初学者更了解其中的内涵，令大家在品尝时更加潇洒和自信，我觉得也是增加了很多情趣和酒话。

对于追求口感的酒鬼来说，威士忌的"泥煤味"，是一种比较重口味的风格，强烈的烟熏气息（你在巧克力里偶尔也能尝到），浓郁又厚重。不同酒厂的泥煤风味还不尽相同，呈现出各种与熏烤有关的风味，比如灰烬、烧焦木头、篝火、熏肉、熏鱼、烤肉、腊肉、碘酒、消毒药水等。

在苏格兰遍地都是泥煤，非常容易取得，即使它的燃烧效率不高，且燃烧过程中烟很大，但是因为实在是太廉价了，所以酒厂在加热蒸馏器和烘烤麦芽时都会使用它。简单来说，泥煤味就是用泥煤熏出来的。非常硬核。泥煤威士忌风味取决于泥煤的组成部分，而泥煤的成分又取决于曾经在那里生长过的植物。例如苏格兰大陆上的毛毯沼泽泥煤主要是由树木构成的，因此与艾雷岛上的泥煤相比，在燃烧时产生更大的愈创木酚和苯酚。而沿海采集的泥煤则经常带有藻类，所以威士忌泥煤有"海"味是很好理解的。威士忌泥煤味道的轻重跟烘烤时间有关，如果你用泥煤烘烤的时间比较短，泥煤味就轻一点。

正所谓萝卜青菜，各有所爱。如果要问俄罗斯人喜欢哪一种酒的话，毋庸置疑，一定是伏特加，说到这种无色无味的伏特加，对于喝酒的新人来说是很容易"上当"的，一些酒鬼会拿伏特加来试探女孩子的酒量，居心叵测！不过也正因它无味的特点，是世界上配搭鱼子酱的最佳伴侣。而威士忌就不一样，尤其是泥煤味的宠儿，拥有先声夺人的"药水"香气，医生护士们对这种熟悉的香味，有着神奇的反应，他们就算不喝酒，也会多嗅几遍，继而点头默认这种"怪味"，正因为这种难得的个性特点，给人们留下了记忆。不过这种味道与食物搭配另当别论。

强烈的个性不一定容易与食物搭配，往往一种霸道能盖过食物的本味，两者没有产生激烈的冲突，也不失为一种口感上的互动。其中的"泥煤怪兽"，我每次饮用时都会产生一种愉悦。认识"泥煤怪兽"是因为我的老友洪潮兄，他曾经是医务人员，而且对每样食物都观察入微，他喜欢与人分享威士忌当中的泥煤ppm，我也从中受益，受益于经常有知己而没有辜负杯中的美妙。所谓的泥煤ppm，本书的作者在不同的篇章中记录了不少的经历。ppm是发芽大麦通过泥煤烟熏烘烤，会吸取的"酚类化合物"，而酚值的高低，也呈现出泥煤烘烤的程度，计算单位是ppm（百万分之一）。ppm值越高，泥煤的含量就越高。在喜欢威士忌的人眼里，威士忌那无可比拟的深度和广度，给威士忌与美食的搭配带来了无数可能。打开我们的想象，威士忌与美食的搭配就不会局限在食物与酒的简单搭配上的。比如一些有经验的厨师，会在食材中添加几滴威士忌，让威士忌的风味激活食材的各种味道，从而给食用者创造一个前所未有的用餐体验。如泥煤味威士忌和生蚝的搭配，艾雷岛威士忌的泥煤味和淡淡烟熏味会让生蚝更甜美。

威士忌的个性特点可以从书本找到很多答案。而说到它所搭配的食物、杯具或者冰、水等，从关系上不算太明显，远远不同于葡萄酒的玩法，比如很多人说红葡萄酒配雪茄，虽然这两者的关系算不上是"郎才女貌"，但也经常让人在不同场合上提起，就算没有体验过的人，不时也会说上两句这样的酒话，算得上是一种传播。但很少有人说到威士忌配雪茄，其实我觉得这两者反而是更加匹配的，当然要看是什么雪茄，配哪一种威士忌。另一种搭配，威士忌加冰或是水对口感有着最直接的影响，一般来说，老酒鬼饮用单一麦芽时从不加任何东西，大不了也是加一滴水而已，而且选用缩口杯，此举能大幅度保留威士忌的香气。高品质的威士忌对杯的要求是苛刻的，皆因不同酒杯能感受不同的效果，通过细品慢尝，有些酒会被发现散发出令人惊喜的味道。其实，威士忌杯子造型和质地跟葡萄酒杯和白兰地杯一样，总是给人带进艺术的欣赏空间，作为一种艺术品而超越了嗅觉与味觉的属性。

酒精的作用总是使得人产生兴奋，仿佛在天马行空的幻觉之中，自己的杯中酒是最香的。弹指之间，转眼已经数十载，无论是在英国苏格兰进修酒店管理，还是在苏格兰的威士忌酒厂里的行踪，常常记忆犹新，好酒也喝了一些，但已物是人非。在我的酒店生涯里，曾有一段时间经常在酒吧与上司聊工作，工作之余，美籍华人老总温先生，总是喜欢喝上一杯带有烟熏味的威士忌，那味道已经成为他生命的一部分，不难看到，这酒激发了他在工作上的灵感。这也是我因为工作的关系，喝过最多威士忌的日子，多年后知道他离世，更加令我回忆当年的味道。不难想象，为什么总有人把一些佳酿命名为"生命之水"，它的含义对每个人都会产生不同的发酵作用。而我当年接触的生命之水，到今天而言，也有很多复杂的变化。法国人命名的Eau de vie是他们老家的干邑，而在伟大的中国文字里，当然少不了本书的主题"威士忌"。

庄臣 / 文

于莹莹 / 译

庄臣，世界美食大使（ihra），法国国际美食会授美食博士，法国国际厨皇会授国际烹饪艺术大师，著名美食专栏作家，现任广东省食文化研究会会长，广州旅游美食形象推广大使。他出生于西关粤剧世家，自小热爱文学和美食。他曾参加央视纪录片《舌尖上的中国》和《穿越海上丝绸之路》的拍摄。他出版了15本食品和饮料书籍，例如《葡萄酒佐餐艺术》，并在《广州日报》的《庄臣食单》栏中发表了4000多篇文章。

PREFACE

In my memory, the first time I got to know whisky was a glass of "WEISHUI", which was whisky mixed with water. That is one way of drinking by whisky fans. In Guangdong, "WEISHUI" also has another meaning of "powerful".

In the 1980s, Hong Kong and Macao played important roles in promoting hotels' reformation in mainland China and building international hotels. Many catering professional terms were named after Cantonese or Hong Kong's and Macao's. I joined an international 5-stars hotel based on Hong Kong management culture at that time. That was when I became attached to and started knowing whisky from "WEISHUI" to whisky with ice. It is quite touched when looking back, as the saying "life in a glass of liquor" goes. The taste of whisky may be vague, but the memories in the taste will keep.

The sunset slanted in the corner of the hotel lobby, where was the lobby bar called "Corner Bar" . The corner always gave you a feeling of mystery. The bar opened on this corner, which also brought a bit of curiosity to people. Faces of foreigners appeared at around 5:30 every afternoon , in my memory.They were mostly from foreign companies in China after reform and opening-up, and they brought in the life style of "happy hour". The tall bar chair could be half sitting and half standing. A hook under the bar counter was convenient for ladies to hang handbags. There was a bell on the bar table for the so-called gentlemen to "ring", which was a signal of paying all the bills. All the British bars had this thing which were really not for playing, because even if it was rung accidentally by children, their parents had to pay. These English style bars are basically the same as later whisky bars. The latter has nothing but more sense of distance.

"ZHAN WEI DI FU LIN" is a simple summary in Chinese of the names of the western hard liquor. Gin, vodka, whisky, ram, brandy, the knowing of these kinds of spirits could also be treated as measuring the drinker's cognitive level of bars.For example, the knowing of cocktails or match of different spiritsand flavours, taste, or even background stories These are memories and talks then spending good times in bar. As a medium, liquor makes people think actively and increases the art of communication between people. Another important measurement is, mature drinkers like whisky, and old drunkards always say that switching from beer to whisky is a signal of growing up. It doesn't make any sense to me, for some drunkards never drink whisky in their lives. Of course, many drunkards say whisky is the most delicious liquor in the world.

I have always objected to calling brandy "foreign wine" in general, but this beautiful misunderstanding could not be changed today. Even children call brandy "foreign wine". But it also led to the separate translation for whisky, and the subsequent subdivision of liquor, in which the subdivision is much more detailed than "foreign wine" (brandy), which is very penetrating, and you will feel the author's work in the book.

The most degraded time of whisky in China was the time of whisky drunk with green tea. Before long this drinking style was very popular in bars for several years. Under the promotion of various business methods, this drinking style caused certain market effect, but it also died out in the end. After a period of cooling-off, people started to notice the existence of "single malt". Chinese people's consumer psychology put whisky under spotlight again due to inside and outside causes. Although it is still a minority, it is not difficult to see the value of the market due to the hype and price rise of the brand. Besides, whisky is mostly spread in the image of gentleman, which, to some extent, makes whisky, this common consumer goods in its country of origin, become enigmatic luxury goods, causing people to follow it blindly, and then start to study it. This book shows us a new view of Islay Whisky's development history, production process and history of distilleries, interprets the nine distilleries and companies on the island, and shares nearly 100 stories of classical brands of Scotch whisky, which lets whisky fans and beginners know more about whisky's connotation, and makes us feel more natural and confident when we taste it. I think it also adds a lot of interests and topics .

For a taste-seeking drunkard, whisky's "peat" is a strong, smoky flavour (you can occasionally taste it in chocolate), rich and heavy. The peat flavour varies from distillery to distillery, presenting a variety of flavours related to smoking and roasting, such as ashes, charred wood, bonfire, bacon, smoked fish, roast meat, bacon, iodine, disinfectant, etc.

Peat is readily available all over Scotland, even if it is not burned efficiently and has a lot of smoke.But it is so cheap that distilleries use it to heat stills and roast malt. Simply put, peat smell is produced by peat smoke. Quite hardcore. The flavour of peat whisky depends on the composition of the peat, which depends on the plants that once grew there. For example, the blanket bog peat in Scotland is mainly made up of trees, so when burned, it produces more guaiacol and phenol than the peat from isles of Islay. While peat from the coast is often laced with algae, it is easy to understand the "sea" flavour of whisky from there. The strength of the whisky peat is related to the roasting duration. If you use peat for a short roasting, the peat taste will be lighter.

Every man has his hobbyhorse. If you ask Russian which kind of liquor they like, no doubt, it must be vodka. Speaking of this colorless and tasteless vodka, it is very easy to "trap" for the new drinkers. Some alcoholic uses vodka to test a girl's drinking capacity, having ulterior motives! But because of its tasteless quality, it is the best caviar accompaniment in the world. Whisky is different, especially it is mud taste's belover, having a sound loaded "medicine" aroma. The doctors and nurses are familiar with this fragrance, and they have a magical reaction for whisky. Even if they don't drink, they will smell a few times more, then nodding as acquiescence of this kind of "taste". This unique character impresses people,but the combination of this taste and the food is another matter.

Strong character is not necessarily easy to match with food. Usually a domineering taste can overwhelm the original taste of food, and the two do not produce a fierce conflict, but also can be regarded as a kind of interaction on the taste. One of them, the "peat monster", was a delight to me every time I drank it. I know the "peat monster" because of my old friend, Chao Hong, who used to be a medical worker and observed every kind of food carefully. He likes to share peat's ppm of whisky with others. I also benefit from it, for having friends frequently and living up to the beauty in the glass. The author of this book, records many experiences of ppm in different chapters. Ppm refers to the "Phenol compound" absorbed by malted barley after being smoked and roasted by peat. The phenol value also shows the degree of peat baking, and the unit of calculation is ppm (Parts Per Million). The higher the ppm is, the higher the peat contents. In the eyes of those who love whisky, the incomparable depth and breadth of whisky bring endless possibilities for the combination of whisky and food. Open our imagination, the collocation of whisky and fine food will not be limited to the simple collocation of food and drinking alcohol. For example, some experienced chefs will add a few drops of whisky to the ingredients, so that the whisky flavour can activate the various flavours of the food, thus creating an unprecedented dining experience for the diners. Like the combination of peat whisky and oysters, the peat and slightly smoky flavours of Islay whisky makes the oysters sweeter.

Many characters of whisky can be found in this book. When it comes to its collocation with food, glass, or ice, water, etc., the relationships are not too obvious, which is far different form wine. For example, a lot of people say red wine matchescigar, though the relationship between the two is not a "perfect match".Still, it has often been mentioned on different occasions, even if someone doesn't have the experience, from time to time he may say something on this, which is a kind of transmission. But very few people talk about whisky with cigars, and I actually think they are a better match, depending on what kind of cigar it is and what kind of whisky it is. On the other hand, whisky mixed with ice or water has the most direct effect on taste. Generally speaking, old drunkards drink single malt without adding anything. A drop of water at most and small end glass for sure, causing it could maximally maintain the fragrance of whisky. High quality whisky is very demanding on the glasses, because different glasses have different effects. Through carefully taste, some liquor can be found to emit a surprising tasting. In fact, the shape and texture of whisky glasses, like wine glasses and brandy glasses, always lead people into the space of art. As a work of art, it goes beyond the smell and taste.

The effect of alcohol always makes people excited, in unrestrained fantasy, and people believe their own glass of drink is the most fragrant one. In the blink of an eye, it has been decades, and it often reminds me of the time when I studied hotel management in Scotland, and in Scotland's whisky distilleries whereabouts. In that time, I drank some fine wine, and now things have been changed. In my hotel career, there was a time when I often talked about work with my superior in bar. After work, Mr. Wen, the Chinese-American boss, always liked to drink a glass of whisky with smoky taste. The taste has become a part of his life, and it was easy to see that this liquor inspired him on his work. It was also the days I drank the most whisky because of my job. Many years later, when I knew that he had passed away, I cherished the memory of the taste of that time even more. It's not hard to imagine why some of the best vintages have been named " the water of life", with a meaning that is different for everyone. And the water of life that I was in contact with then, till today, also has a lot of complex changes. "Eau de vie" named by French is the cognac of their hometown, while putting into the great Chinese word, leads to the theme of "whisky" of this book.

Writed by Johnson Wong

Translated by Yingying Yu

Johnson Wong, World Cuisine Ambassador(ihra), Senior Gourmet of Chaine Des Rotisseurs, Master of the Art of Fine Dining of Les Amis d'Escoffies Society.Inc, famous food columnist, President of Guangdong Food Culture Research Council , Guangzhou Tourism & Food Image Promotion Ambassador. He was born in a family of Xiguan Cantonese Opera,and he loves literature and food since his childhood. He once participated in the shooting of CCTV documentaries *A Bite of China* and *Crossing the Maritime Silk Road*. He has published 15 food & beverage books such as *Art of Wine Accompaniment*, and published more than 4000 articles in the column of "*Johnson Recipe*" of *Guangzhou Daily*.

目 录
CONTENTS

1.关于艾雷岛

About Islay

被誉为"赫布里底群岛女王"的艾雷岛，是内赫布里底群岛最南端的岛屿，也是苏格兰第五大岛屿，坐拥620平方公里资源，南北纵贯40公里，东西横跨24公里。

它位于阿盖尔—比特议会区内，距朱拉岛西南仅几百米——两者被艾雷海峡隔开，位于爱尔兰以北40公里。

海拔491米的贝恩·比格峰是艾雷岛的最高点。

Islay, known as "The Queen of the Hebrides", is the southernmost island of the Inner Hebrides and the fifth-largest Scottish island, with an area of 620 square kilometres, running about 40 kilometres from north to south, and 24 kilometres from east to west.

It lies in Argyll and Bute Council, just a few hundred metres south-west of Jura, separated by the Sound of Islay, and around 40 kilometres north of the Ireland.

The highest point of Islay is Beinn Bheigeir summit with 491 meters.

埃伦港 Port Ellen（照片由作者提供 Photo by author）

气候

受墨西哥湾暖流影响，艾雷岛气候温和，夏季平均气温为16℃。然而，尽管岛上雪和霜冻相对罕见，它仍然被来自大西洋时速超150公里的强风影响着。

艾雷岛年降雨量125厘米，是伦敦的两倍。4月、5月和6月是最干旱的月份，而最潮湿的月份是1月和10月。

人口

该岛行政首府是位于岛中心的波摩，位于洛欣达尔湾的海岸线上。埃伦港是最大的镇，夏洛特港和波特纳黑文是另外两个主要定居点。19世纪初，该岛人口已有约1.8万，如今仅3000多人。

在这里，主要的经济活动是农业、麦芽威士忌蒸馏业和旅游业。

Climate

The climate is mild and ameliorated by the Gulf Stream, with average summer temperature of 16°C. However, the island is influenced by strong winds from the Atlantic at speeds of more than 150 km/h, although snow and frosts are relatively rare on the island.

Islay receives 125 cm of rain every year, twice more than London. April, May and June are the driest months and January and October are the wettest.

Population

The island's capital is Bowmore, located at the heart of the island, on the shore of Loch Indaal. Port Ellen is the biggest town and Port Charlotte and Portnahaven are the other main settlements.The population of the island was around 18,000 at the beginning of the 19th century, but today it is just over 3,000.

The main economic activities are agriculture, malt whisky distillation and tourism.

阿斯凯格港渡轮 Port Askaig ferry（照片由作者提供 Photo by author）

运输

岛上有两个渡口：

埃伦港——可航行到肯纳奎格（位于苏格兰大陆）。

阿斯凯格港——渡运到肯纳奎格（大陆）、奥本（大陆）以及费林（朱拉岛上）。

埃伦港附近还有一个机场，有航班到格拉斯哥、奥本以及科伦赛。

旅游魅力

威士忌显然是艾雷岛旅游业的核心吸引力之一，所有的酒厂都提供参观及试饮。但同时，艾雷岛也是许多鸟类的栖息地，比如越冬种群格陵兰白额雁和白颊黑雁。这里常年是观鸟爱好者们的热门目的地。

以下是岛上其他几个可以造访的地方：

美国人纪念碑——位于艾雷岛西南角的奥之岬，这个纪念碑纪念着在1918年两次海难中失去的几百条生命。

Transport

There are two ferry ports on the island:

Port Ellen – ferry to Kennacraig (Mainland);

Port Askaig – ferry to Kennacraig (Mainland), Oban (Mainland) and Feolin (Island of Jura).

There is also a small airport near Port Ellen, with flights to Glasgow, Oban and Colonsay.

Attractions

Whisky is obviously a key attraction of Islay, with all distilleries offering tours and tastings. But also Islay is home to many bird species such as the wintering populations of Greenland white-fronted and Barnacle goose, and is a popular destination throughout the year for birdwatchers.

Here are some other places that you can visit on the island:

The American Monument — Located on the Oa in Islay's south-west corner, the American Monument commemorates the hundreds of lives lost following two shipwrecks in 1918.

鸣沙滩 The Singing Sands Beach（照片由作者提供 Photo by author）

鸣沙滩——得名于，当你在这里的沙子上拖动鞋底，沙滩会发出鸣叫声。沙滩位于奥之半岛，就在灯塔后面。

卡里·发达灯塔——独特的灯塔，由华特·费德里克·坎贝尔于1832年为纪念埃莲娜·坎贝尔女士而委托建造，位于埃伦港外。

奇戴顿十字架——公元800年左右建成的一个天主教十字架，高2.65米，宽1.32米。位于奇戴顿老教区教堂内，接近艾雷岛东南角。

菲拉根城堡——近一千年前就在这里，"诸岛领主"统治着艾雷岛。它位于艾雷岛北部的菲拉根湖，涵盖两座岛屿，其中一座通过一条步道与大陆相连。三座建筑的废墟至今仍能看到，而考古学家也曾挖掘出城堡存在过加固建筑、墓园和议事厅的证据。

The Singing Sands Beach — Its name comes from the fact that if you move the soles of your shoes over the sand it will start to sing. The beach is located on the Oa peninsula just behind lighthouse.

Carraig Fhada Lighthouse — The unique lighthouse was commissioned to build in 1832 by Walter Frederick Campbell in memory of Lady Ellenor Campbell. The Lighthouse lies just outside of Port Ellen.

The Kildalton Cross — A Christian cross erected around 800 AD, which stands 2.65 m high and spans 1.32 m across. The cross is located in the Old Parish Church of Kildalton near the south-east tip of the island.

Finlaggan Castle — The site from where, nearly 1000 years ago, the "Lords of the Isles" controlled Islay. It is located on the Loch Finlaggan, on the north of the island. It comprises two islands, one of which is linked to the mainland by a walkway. The ruins of the three buildings are still visible today and archaeological digs uncovered the evidence of fortified buildings, graveyards and a council chamber.

艾雷屋广场——由华特·坎贝尔于18世纪90年代建成，这个广场就在布里真德外。它原本用来做艾雷屋雇员宿舍以及畜舍，但是今天它成为很多当地生意的大本营，例如木匠工作室、玻璃工作室和一家啤酒厂。

艾雷岛毛纺厂——建立于1883年的艾雷岛毛纺厂生产短裙、毯子、围巾和帽子已经超过100年。他们的设计被用在一些知名电影中，比如《勇敢的心》和《阿甘正传》。毛纺厂在布里真德外大约1.5公里处。

艾雷岛生活博物馆——1977年开业并拥有庞大的馆藏，部分馆藏品的年代可追溯至公元前8000年。博物馆位于夏洛特港。

马希尔湾（又译：玛吉湾）——艾雷岛上最美的海滩之一。马希尔湾在艾雷岛西海岸，靠近齐侯门，金色的沙滩绵延伸展超过一英里。

马希尔湾 Machir Bay（照片由作者提供 Photo by author）

Islay House Square — Established by Walter Campbell in the 1790s, Islay House Square lies just outside Bridgend. Its original use was as accommodation for employees and stables of Islay House, but today it is home to many local businesses, such as brewery carpenter's and glassmaker's workshops.

Islay Woollen Mill — Founded in 1883, the Islay Woollen Mill produces kilts, rugs, scarves and caps for more than 100 years. Its designs have been used in major films such as *Braveheart* and *Forrest Gump*. The Woollen Mill is located approximately 1.5 km outside Bridgend.

Museum of Islay Life — The Museum of Islay Life opened in 1977 and has a huge collection of items, some of which date to 8000 BC. The museum is based in Port Charlotte.

Machir Bay — One of the most beautiful beaches on Islay. Machir Bay on the island's west coast, near Kilchoman, stretches for more than one mile with golden sand.

2.艾雷岛的历史
The History of Islay

　　艾雷岛最早的居民，是更新世冰盖消退后的中石器时代到来的游牧狩猎采集者。

　　这个时期的谋生活动可能是季节性的，可以利用沿海的丰富食材资源。遗迹表明，这一时期的建筑是在沙子里挖出的简陋居所。

　　随着气候的改善，第一批农民来到艾雷岛肥沃的土地上定居下来。

　　可惜的是，关于铁器时代的艾雷岛，即公元前1000年中期至公元1000年中期这段时期的研究很少。这个时期内留下的最好的建筑物是位于布里真德东南的鼻桥堡垒。这个占地375平方米的要塞雄踞一块凸出的峭壁，周围景色一览无余。

洛欣达尔湾 Loch Indaal（照片由作者提供 Photo by author）

格姆湖 Loch Gorm（照片由作者提供 Photo by author）

The earliest settlers on Islay were nomadic hunter-gatherers who arrived during the Mesolithic period after the retreat of the Pleistocene ice caps.

Occupation at this time may have been seasonal, taking advantage of the rich coastal food resources. Remains indicate that the buildings of this period were rough shelters dug into the sand.

With climate improvement the first farmers come and settle on the fertile lands of Islay.

Iron Age on Islay, extending from the mid first millennium BC to the mid first millennium AD is, unfortunately, poorly researched. Probably, the best structure from that period on the island is Dun Nosebridge, southeast of Bridgend. This 375 square metres fort occupies a prominent crag and has commanding views of the surrounding landscape.

波摩圆形教堂 Round Church of Bowmore
（照片由作者提供 Photo by author）

公元500年以后，基督教传入艾雷岛，在一系列石刻十字架和教堂中留下了自己的印记。圣高隆的传教士建立了许多早期的基督教场所，并建立了小型教堂。

在后来信奉基督教的挪威殖民者时期，岛上的每个行政区都修建了自己的教堂和墓地。墓地里竖起纪念性十字架，有的至今还能看到。奇戴顿十字架是苏格兰唯一现存的完整的凯尔特人十字架。这个十字架高2.65米，年代可追溯至公元800年。它也许是由一个来自爱奥那岛的雕刻家用当地产的青石雕刻而成的。它正面雕刻的圣经场景，包含了圣母和圣婴。

公元6世纪，艾雷岛连同附近大部分大陆以及周边岛屿都属于盖尔人的达尔里阿达王国，这个王国与爱尔兰有着千丝万缕的联系。

公元9世纪，挪威人来到海岸上，从开始的入侵者，变成后来的商人和定居者，并建立起"诸岛之国"，后来这个王国在挪威统一之后成为挪威"王冠"上的一部分。在接下来的4个多世纪里，这个王国都处于挪威大本营的控制之下。

From the period, after 500 AD, onwards, Christianity came to Islay, leaving its mark on a series of carved stone crosses, as well as churches. Columba's missionaries established many early Christian sites and built small churches.

During the period of the Norse colonists, who later on adopted the Christian religion, each district on Islay built its own church and burial ground. Commemorative High crosses were raised on the burial grounds and some of them can still be seen today. Kildalton High Cross is the only surviving complete Celtic High Cross in Scotland. The cross is 2.65 metres tall and can be dated back as far as 800 AD. It was carved, probably, by a sculptor from Iona, from the local blue stone. The biblical scenes, on the front, include the Virgin and the Christ child.

By the 6th century AD Islay, along with much of the nearby mainland and adjacent islands, lay within the Gaelic Kingdom of Dál Riata with strong links to Ireland.

At the 9th century AD, Norse had arrived on the shores, first as raiders but later as traders and settlers and established the Kingdom of the Isles, which became part of the crown of Norway following Norwegian unification. For the next four centuries and more this kingdom was under the control of rulers of mostly Norse origin.

　　最终，挪威侵入者和当地家族通婚，在这样的社会环境中出现了很多有权势的人，其中最重要的一个人就是苏摩烈。他利用马恩岛的奥拉夫国王驾崩的时机，在1156年入侵了南赫布里底群岛并成为那里的统治者。接着，他在爱尔兰活动并与苏格兰国王对立。在他死后，他的儿子雷纳德继承了他的地位，自封为诸岛之王和阿盖尔领主。在雷纳德之后，他儿子唐纳德，也就是唐纳德家族的建立者，继承了艾雷岛王国。

　　随着对苏格兰的战败，根据1266年签订的《珀斯条约》，诸岛的统治权被割让给了苏格兰王室。

　　安格斯·奥格领导下的麦克唐纳家族，在苏格兰独立战争中支持了罗伯特·布鲁斯，新国王罗伯特在14世纪初将土地还给了他们。安格斯的儿子约翰是第一个授予自己"诸岛领主"头衔的人。

　　领主权力的基础集中在艾雷岛上的菲拉根。

　　1493年，当最后一位也是最有野心的领主约翰二世与英国国王结盟，对抗苏格兰国王詹姆斯三世，并在与詹姆斯四世的战斗中被击败时，统治结束了。

　　在随后的政治真空中，发生了许多叛乱，直到国王詹姆斯四世将艾雷岛上的土地归还给一位麦克唐纳家族的成员——阿德纳默亨半岛的约翰，秩序才得以恢复。在约翰的统治下，一个新的审判系统创立了，土地估价被贯彻，教会被改革。

In time, the Norse invaders married into local families. From this society there came a number of powerful men, of which Somerled was the most important. Somerled took advantage of the death of King Olaf of Man, invaded the southern Hebridean Isles in 1156 and established himself there as the ruler. He continued to campaign in Ireland and against the King of Scotland. Following his death, his son Ranald took his place, naming himself as King of the Isles and Lord of Argyll. After him, his son Donald, founder of the Clan Donald, inherited the Kingdom of Islay.

Following defeat in battles against the Scots, the rule of the isles was ceded to the Scottish crown under the *Treaty of Perth,* signed in 1266.

After MacDonalds under Angus Og, supported Robert Bruce in the Scottish Wars of Independence, the new King Robert granted them back lands in the early 14th century. Angus's son John was the first to give himself the title of Lord of the Isles.

The power base of the lordship was centred at Finlaggan on Islay.

The lordship was ended in 1493 when the last and most ambitious of the Lords, John II struck an alliance with the English King against King James III of Scotland and was defeated in battle by James IV.

In the political vacuum which followed, there were numerous rebellions and order was not restored until King James IV returned lands on Islay to John of Ardnamurchan, a MacDonald. Under his rule, a new court system was instigated, land valuations were carried out and the church was reformed.

在约翰死后，艾雷岛的管理权移交给了考德的坎贝尔，后来在1528年移交给了阿盖尔伯爵亚历山大·马克伊安。在争执后，最终，1542年，大部分土地落入了詹姆斯五世国王手中。

然而争斗还在小规模地持续着，最终在格林纳特湖爆发的麦克唐纳家族和麦克林斯家族针对林斯所有权的战斗中达到顶峰。

1608年，麦克唐纳家族对苏格兰宗教改革表现出的敌意导致苏格兰—英格兰王室发动了一次针对他们的远征。1614年，国王把艾雷岛交给了考德的约翰·坎贝尔爵士，作为后者承诺平息艾雷岛局势的回报。在坎贝尔家族的影响下，阿盖尔郡治安官树立起权威。在坎贝尔家族掌控了司法辖区后，艾雷—朱拉的区划逐渐消失。

坎贝尔一家的领主们大部分是外居领主，尽管他们有一些改进农业和引进新工业的尝试，但艾雷岛的经济状况在17世纪大部分时间里都萎靡不振。

1726年，一位富有的烟草大亨、国会议员，来自肖菲尔德的丹尼尔·坎贝尔买下了艾雷岛。随着所有权的变化，艾雷岛开始复苏。除了改良耕作方法外，这位新主人还引进了亚麻种植、工坊和织工来从事亚麻布的生产。

丹尼维哥城堡的废墟 The ruins of Dunyvaig Castle
（照片由作者提供 Photo by author）

When John of Ardnamurchan died, the administration of Islay was passed to Campbell of Cawdor and latterly in 1528, to the Earl of Argyll, Alexander MacIan. After disagreement, at the end, much of the lands fell to King James V in 1542.

Feuding continued on a smaller scale, however, culminating with a battle at Loch Gruinart between the MacDonalds and MacLeans over the ownership of the Rhinns.

In 1608, MacDonald's hostility to the Scottish Reformation led the Scottish-English crown to mount an expedition to subdue them. In 1614 the crown handed Islay to Sir John Campbell of Cawdor, in return for an undertaking to pacify it. Under Campbell's influence, authority was established under the sheriff of Argyll. With inherited Campbell control of the sheriffdom, the provincial identity of Islay-Jura faded away.

The Campbells was mostly absentee lairds and despite some attempts to improve farming and introduce new industry, Islay languished for much of the 17th century.

In 1726, Islay was purchased by Daniel Campbell of Shawfield, a wealthy tobacco baron and member of Parliament. With the change of ownership, Islay began to revive. Together with improving farming methods, the new proprietor also introduced flax cultivation, mills and weavers to engage in linen production.

埃伦港战争纪念馆 Port Ellen War Memorial
（照片由作者提供 Photo by author）

他的继承人，小丹尼尔，肩负起更长远的发展责任，包括建设波摩的村庄、发展渔业以及提供学校。

1899年，经由一项苏格兰当地政府法案，沿辖区边界正式建立起郡县，艾雷岛于是成了阿盖尔郡的一部分。

第一次世界大战期间的1918年，两艘运兵舰在仅几个月内相继在艾雷岛附近沉没。运兵舰"SS图斯卡尼亚号"在2月5日被德国潜艇UB——77用鱼雷击沉，超160人丧生，此船至今长眠在奥之岬以西6.4公里处的深海里。10月6日，"HMS奥特朗托号"在波涛汹涌的海面与"HMS克什米尔号"相撞，并于马希尔湾附近海岸失事，夺走了431条生命。

美国红十字会在奥之岬海岸修建了一座纪念碑以悼念这两艘沉没的舰船。

二战期间，皇家空军在格兰尼格黛尔修建了一座机场，这个机场后来成为艾雷岛的民用机场。

His successor, Daniel the Younger, was responsible for further improvements, including construction of the village of Bowmore, the development of the fishing industry and the provision of schools.

In 1899, counties were formally created, on shrieval boundaries, by a Scottish Local Government Act, Islay therefore became part of the County of Argyll.

During World War I, in 1918, two troopships sunk near Islay within a few months. The troopship *SS Tuscania* was torpedoed by German submarine UB-77 on February 5th with the loss of over 160 lives and now lies in deep water 6.4 kilometres west of the Mull of Oa. On October 6th *HMS Otranto* was involved in a collision with *HMS Kashmir* in heavy seas and wrecked on the shore near Machir Bay with a total loss of 431 lives.

American Red Cross built a monument on the coast of The Oa to commemorate the sinking of these two ships.

During World War II, the RAF (Royal Air Force) built an airfield at Glenegedale which later became the civil airport for Islay.

3.艾雷岛威士忌的历史

The History of Islay Whisky

　　人们相信，最早是由爱尔兰僧侣在14世纪早期把蒸馏酒技术传入艾雷岛。他们发现这是一个完美适于生产"生命之水"的岛屿，泥煤取之不尽，湖泊以及河流充盈着纯净的软水，当地佃农种植着现代大麦作物的祖先。

　　在早期，威士忌蒸馏都是在无执照的小酒馆公然进行，直到1644年消费税法案出台，对威士忌征税。这迫使蒸馏厂搬到遥远的峡谷和洞穴，以免被发现，甚至第一个税务官直到1797年才敢来。1777年，齐侯门教区的约翰·麦克利什牧师报告说："我们这整个岛上连一个消费税官员都没有，因此，这里的威士忌产量非常大，并且随着过量饮用这种液体而带来的罪恶，也是显而易见的。"

　　奥之岬半岛曾因非法蒸馏威士忌而闻名，在克拉格布斯、哥尔、下基勒扬和斯奇尼斯摩均发现了（非法）蒸馏器。在岛上的其他地方比如布里真德、达尔、洛西特、暮林据、奥肯特摩和塔兰特也都有蒸馏器。

　　在前些年艾雷岛出产的威士忌中，约95%被用在所有著名的混合威士忌中，比如百龄坛、黑白、芝华士、帝王、约翰尼·沃克、白马，不胜枚举。

威士忌桶 Whisky Casks
（照片由作者提供 Photo by author）

It is believed that during the early fourteenth century, the Irish monks first introduced the art of distillation to Islay. They found an island perfectly suited for the production of Uisge Beathe (water of life). Unlimited supplies of peat, lochs and rivers filled with pure soft water and the local crofters grew the fore-runner of the modern barley.

In the early days, distilling was carried out openly in black shebeens until the introduction of the Excise Act when a tax was levied on whisky, in 1644. This forced the distillers to move into the remote glens and caves to avoid detection, even the first inspector didn't dare to come until 1797. In 1777 the Reverend John McLeish of Kilchoman Parish reported that, "We have not an excise officer on the whole island. The quantity therefore, of whisky made here is very great and the evil that follows drinking to excess of this liquor, is very visible on the island."

The Mull of Oa, peninsula, was well known for illigal distilling, with stills found at Cragabus, Goil, Lower Killeyan and Stremnishmore. There were also stills in other parts of the island as at Bridgend, Dail, Lossit, Mulindry, Octomore and Tallant.

In former years around 95% of the whisky produced on Islay was used in the making of all the famous blends, i.e. Ballantines, Black & White, Chivas Regal, Dewar's, Johnny Walker, White Horse; the list is endless.

艾雷岛上没落的威士忌酒厂

在以前，艾雷岛的酿酒厂比现在多得多。在过去200年里，酿酒厂开张—倒闭—重新开张。艾雷岛最早的合法酿酒厂的记录，是在1779年的波摩，当时有多达23家酿酒厂在运营。

一些没落的酒厂如下：

阿丹尼斯托酒厂 (1837—1852)，1853年被归入拉弗格。

阿德摩酒厂 (1817—1835)，1837年由拉加维林接管。

巴利格兰特酒厂 (1818—1821)。

布里真德酒厂 (1818—1822)，遗址现存于一处村庄中。

德尔酒厂 (1814—1830)，遗址位于布里真德和阿斯凯格港之间。

自由港酒厂 (1847—1847)。

格兰纳维林酒厂 (1827—1832)。

基拉罗酒厂 (1760—1818)。

洛欣达尔湾酒厂 (1829—1929)，在布鲁赫拉迪附近。

洛西特酒厂 (1826—1867)，遗址现存于一座村庄附近。

麦芽工坊酒厂 (1908—1962)，已与拉加维林合并。

暮林据酒厂 (1826—1827)，遗址位于捏利比溪和拉根河的交汇处。

Islay's "lost" Whisky Distilleries

In historic times Islay had much more distilleries than nowadays. In the last 200 years distilleries have started, closed and reopened. The oldest record of a legal distillery on the island of Islay refers to Bowmore in 1779, and at one time there were up to 23 distilleries in operation.

Some of the "lost" distilleries of Islay are:

Ardenistle Distillery (1837—1852), subsumed by Laphroaig 1853.

Ardmore Distillery (1817—1835), taken over by Lagavulin 1837.

Ballygrant Distillery (1818—1821).

Bridgend Distillery (1818—1822), ruins in village.

Daill Distillery (1814—1830), ruins between Bridgend and Port Askaig.

Freeport Distillery (1847—1847).

Glenavullen Distillery (1827—1832).

Killarow Distillery (1760—1818).

Lochindaal Distillery (1829—1929), near Bruichladdich.

Lossit Distillery (1826—1867), ruins near village.

Malt Mill Distillery (1908—1962), merged with Lagavulin.

Mulindry Distillery (1826—1827), ruins on the junction of the Neriby Burn and the River Laggan.

牛顿酒厂 (1819—1837)，遗址位于布里真德和阿斯凯格港之间。

奥肯特摩酒厂 (1816—1854)，遗址位于夏洛特港附近。.

奥托维林酒厂 (1816—1819)。

埃伦港酒厂 (1825—1929 / 1967—1983)，已经改建成了麦芽厂。

斯卡拉巴斯酒厂 (1817—1818)。

塔兰特酒厂 (1821—1852)，位于波摩以南。

2017年10月，帝亚吉欧宣布，他们将在2020年让标志性的埃伦港酒厂起死回生。

布里真德 Bridgend（照片由作者提供 Photo by author）

艾雷岛的山丘 Hills of Islay（照片由作者提供 Photo by author）

Newton Distillery (1819—1837), ruins between Bridgend and Port Askaig.

Octomore Distillery (1816—1854), ruins near Port Charlotte.

Octovullin Distillery (1816—1819).

Port Ellen Distillery (1825—1929 / 1967—1983), converted to Maltings complex.

Scarrabus Distillery (1817—1818).

Tallant Distillery (1821—1852), south of Bowmore.

In October 2017 Diageo announced that they would bring back to life the iconic Port Ellen Distillery, by 2020.

4.艾雷岛的威士忌

Islay Whisky

艾雷岛的威士忌通常与富有泥煤味的单一麦芽威士忌联系在一起。艾雷岛上单一麦芽威士忌的这一核心特质归因于用来酿酒的大麦的泥煤度。它们还带有碘酒、海藻和盐的味道，许多人将之描述为"药水"味。

阿贝、拉弗格以及拉加维林，这三大酒厂位于艾雷岛南部海岸，已经世界闻名，生产一些优质的泥煤质单一麦芽威士忌。同样也生产重泥煤味威士忌的卡尔里拉则位于岛的北边。

也有一些泥煤味不那么重的威士忌，比如位于该岛以北的布纳哈本，它出产的威士忌酒体就轻很多。还有坐落于艾雷岛行政首府的波摩，它出产的威士忌平衡度非常好，泥煤味强度适中。

布鲁赫拉迪酿酒厂以其在生产单一麦芽威士忌时的实验立场而闻名，其中一个例子就是重度泥煤威士忌奥肯特摩（含量高达309ppm）。

齐侯门的农庄酒厂自2005年末投产，通常也用泥煤，但是不到南部三大酒厂那种程度。它所处的地理位置也不同于其他7个酒厂，并不在海岸边。

随着阿德纳霍酒厂已于2018年投产，岛上有9家活跃的酒厂。每家酒厂都生产独一无二、与众不同的威士忌。我敢说，同厂牌中堪称独特的威士忌一定不止一种。

艾雷岛景观 Islay landscape（照片由作者提供 Photo by author）

Islay whisky is usually associated with peaty single malts. This central characteristic of the Islay malts is ascribed to the peating levels of the barley used in distillation. They also possess notes of iodine, seaweed and salt. Many describe this as a "medicinal" flavour.

Ardbeg, Laphroaig and Lagavulin, the three big distilleries on the south coast, which have become world famous, produce some excellent peaty single malt whiskies. Also producing a strongly peated whisky, Caol Ila is located on the northern side of the island.

There are also some less peaty drams, like Bunnahabhain, which sits to the north of the isle and produces much lighter whiskies and Bowmore, in the island's administrative capital, produces a whisky which is well balanced, using a medium strong peating level.

Bruichladdich distillery is well known for their experimental stance when producing single malt, for example, Octomore which is heavily peated (up to 309 ppm).

The farm-distillery of Kilchoman, started production in late 2005, is usually peated, but not to the same level as the three distilleries on the south coast. Its location is unlike the other seven distilleries, as it is not situated on the coast.

With Ardnahoe who just started production in 2018, there are nine distilleries currently active on the island. Each of them produces a unique, distinctive whisky. More than one, I dare to say.

酒厂前面的威士忌桶 Whisky casks in front of the distillery
（照片由作者提供 Photo by author）

5.艾雷岛威士忌——酒厂和公司

Islay Whiskies
—Distilleries and Companies

阿贝 / Ardbeg

　　这个名字来源于苏格兰盖尔语的"Àrd Beag"，意思是"小高地"。

　　阿贝酒厂自从1798年就开始生产威士忌，而商业化生产始于1815年。像大多数苏格兰威士忌酒厂一样，在大部分历史时间里，它生产的威士忌是用于混合威士忌，而不是作为单一麦芽威士忌而存在。到1887年，酒厂每年的产量是250,000加仑（110万升）威士忌。1911年，阿贝这个名字被注册为商标。

　　在阿贝被希兰·沃克所拥有期间，生产于1981年停顿，但在1989年有限地恢复，一直维持在低水平，直到1991年再次关闭。这家酿酒厂被法国酩悦·轩尼诗—路易·威登集团(LVMH)旗下的格兰杰收购，1997年恢复生产。

　　这家酿酒厂由埃德·多德森重新开张，后来交给斯图尔特·汤姆森，汤姆森在1997年至2006年间经营这家酒厂。迈克尔·海德斯（昵称"米奇"），一个土生土长的艾雷岛人，曾在朱拉酒厂担任经理，几年前就曾在阿贝工作过，在2007年接手了阿贝酒厂。

　　从2011年到2014年，阿贝在国际空间站进行了第一次太空威士忌实验。

阿贝酒厂前 In front of the Ardbeg Distillery
（照片由作者提供 Photo by author）

其中一座酒厂建筑
One of the distillery
buildings（照片由作者提
供 Photo by author）

The name is derived from the Scottish Gaelic: Àrd Beag, meaning *Little Height*.

The Ardbeg Distillery has been producing whisky since 1798, and began commercial production in 1815. Like most Scottish distilleries, for most of its history, its whisky was produced for use in blended whisky, rather than as a single malt. By 1887 the distillery produced 250,000 gallons (1.1 million litres) of whisky per year. In 1911 the name Ardbeg was registered as trademark.

During the period when Ardbeg was owned by Hiram Walker, production was halted in 1981, but resumed on a limited basis in 1989 and continued at a low level till 1991 when it was closed again. The distillery was bought by Glenmorangie plc (owned by the French company LVMH) with production resuming in 1997.

The distillery was reopened by Ed Dodson and handed over to Stuart Thomson, who managed it from 1997 to 2006. Michael "Mickey" Heads, an Islay native and former manager at Jura Distillery who had worked at Ardbeg years earlier, took over in 2007.

From 2011 till 2014, Ardbeg conducted the first whisky experiment in space on the International Space Station.

旧的威士忌蒸馏器 Old whisky still
（照片由作者提供 Photo by author）

阿贝10年 / Ardbeg 10 Years Old

46%ALC./vol.

闻香	泥煤味中注入了柠檬和青柠、黑巧克力、薄荷脑、胡椒、柏油绳、石墨、熏鱼和脆培根的气味。
味觉	一记泥煤和咸味的重击，带着强烈的柠檬和青柠味、胡椒和肉桂味的太妃糖味道。还有香蕉、醋栗、意式浓咖啡和焦油烟的味道。
余味	烟熏味悠长，带着意式黑咖啡、茴香、烤杏仁、少量软大麦和新鲜梨的韵味。
酒桶故事	在美国橡木波本桶中熟成。

Ardbeg

46%ALC./vol.	
NOSE	Peat infused with lemon and lime, dark chocolate, menthol and pepper, tarry ropes, graphite, smoked fish and crispy bacon.
TASTE	A peaty, briny punch with tangy lemon and lime, pepper and cinnamon-spiced toffee. Bananas, currants, espresso and tarry smoke.
FINISH	Long and smoky with tarry espresso, aniseed, toasted almonds and traces of soft barley and fresh pear.
CASK STORY	*Matured in ex-Bourbon American Oak casks.*

阿贝奥之岬 / Ardbeg An Oa

46.6%ALC./vol.	
闻香	如同燃烧的苹果木般，醇厚丰满且有烟熏感。多汁的水果，比如桃子和香蕉。还带有烟熏香草、薄荷太妃糖和一些温和的麦芽饼干的气味。
味觉	柔滑如奶油般的口感引入庞大的糖浆甜味中，牛奶巧克力风味，有一点烟熏茶叶、雪茄的烟雾和烤洋蓟的味道。
余味	余韵悠长、诱人，柔和而强烈，带有八角、山核桃和淡淡的烟味。
酒桶故事	在阿贝酒厂的法国橡木收集桶中创作而成，使用了来自几个不同类型酒桶中的威士忌。

Ardbeg

46.6%ALC./vol.	
NOSE	Rounded and smoky like burning applewood. Juicy fruits, such as peach and banana. Hints of smoked herbs, mint toffee and some gentle malty biscuit notes.
TASTE	A smooth, creamy texture leads into a huge syrupy sweetness, flavours of milk chocolate, smoky tea leaves, some cigar smoke and grilled artichokes.
FINISH	Lingering, seductive, gentle yet intense, with flavours of aniseed, hickory and subtle smoke.
CASK STORY	*Created in Ardbeg's French oak Gathering Vat using whisky from several cask types.*

阿贝奥格阿黛 / Ardbeg Uigeadail

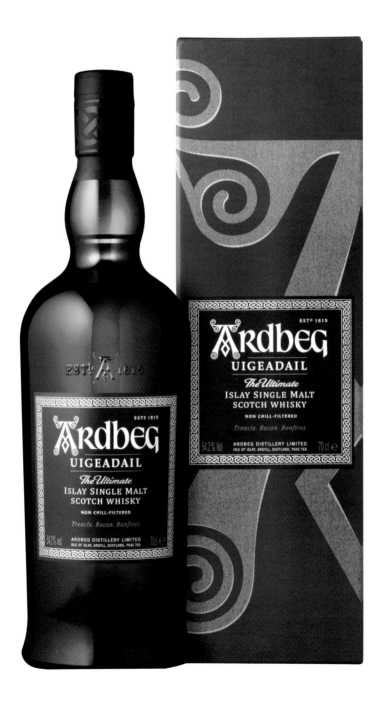

54.2%ALC./vol.

闻香	温暖的圣诞蛋糕、闷烧的煤火气息与皮革、糖浆太妃糖和巧克力葡萄干味融合在一起。加水会带出燃烧的圣诞布丁、柏油烟、柴油、开花醋栗、核桃面包和摩卡浓缩咖啡的香气。
味觉	丰富、油感和黏附感——冬季香料的香味在口中爆发出来。接着发展为涂抹蜂蜜的烟熏食品和糖浆的味道。烟熏味的构建仿佛一支上等的蒙特克里斯托雪茄。
余味	令人惊叹的悠长和浓郁黏稠，带着绵长的葡萄干味、深沉的摩卡调和芬芳的烟熏味。
酒桶故事	熟成于欧罗索雪利桶以及波本桶中。

Ardbeg

54.2%ALC./vol.	
NOSE	Warm Christmas cake, smouldering coal fires. Leather, treacle toffees and chocolate raisins. Adding water brings forth a fired Christmas pudding, tarry smoke, diesel engine oil, flowering currants, walnut bread and mocha espresso.
TASTE	Rich, oily and mouthcoating—winter spices explode. Honey-glazed smoked food and treacle develops. Smoke builds like a fine Montecristo cigar.
FINISH	Amazingly long and chewy with lingering raisiny, deep mocha tones and aromatic smoke.
CASK STORY	*Matured in ex-Oloroso sherry and ex-Bourbon casks.*

阿贝科里弗雷肯（又名：阿贝旋涡）/
Ardbeg Corryvreckan

57.1%ALC./vol.

闻香	令人陶醉,强而有力——木榴油、黑巧克力、黑醋栗和黑砂糖气味;饱满的樱桃和松针香气;加入一点清水,辣椒腌制的牛排、糖浆、丁香和蓝莓的香气便浮现出来。
味觉	犹如胡椒、泥煤味组成的深邃的激流旋涡——带着爽口的海藻、意式特浓咖啡、黑色水果以及八角茴香的韵味。
余味	回味悠长而深沉,留下强有力的巧克力包裹的樱桃、焦糖咖啡和胡椒酱的尾韵。
酒桶故事	熟成于新的法国橡木桶以及波本桶中。

Ardbeg

57.1%ALC./vol.	
NOSE	Heady, intense and powerful—creosote, waxy dark chocolate, blackcurrants and muscovado sugar. Plump cherries and pine needles. A splash of water and cayenne peppered steak, treacle, cloves and blueberries rise to the surface.
TASTE	Swirling torrents of deep, peppery, peaty taste—crispy seaweed, espresso coffee, dark fruits and star anise.
FINISH	Long, deep and remaining powerful with chocolate-coated cherries, tarry coffee and pepper sauce.
CASK STORY	*Matured in Virgin French Oak and ex-Bourbon casks.*

波摩 / Bowmore

　　波摩是艾雷岛上最古老的酒厂，也是世界上最古老的威士忌熟成仓库——1号仓库的所在地。正是在这个仓库里，波摩威士忌得到大师们精心酿制，已有近240年的历史。

　　第一次关于波摩蒸馏厂的记载可追溯到1779年，但如果你和岛上一些年长的居民好好交谈几句，他们可能会私下里告诉你，这里在更早些时候就开始蒸馏酒了。

　　毫无疑问，波摩的创始人，商人大卫·辛普森早在1766年就买下了这块土地。在这期间到底发生了什么，全靠人们猜测。

　　谈到威士忌的传承，波摩酒厂是无可比拟的。他们仍世代相传着早期苏格兰威士忌精雕细琢的传统、技艺和专业技能。

　　以下是波摩蒸馏厂的一些重要年份：

　　1837年，一对格拉斯哥的双胞胎兄弟威廉·穆特和詹姆斯·穆特从辛普森家族手中收购了波摩蒸馏厂。他们用雄厚的家底建成了从拉根河取水的汲水装置。

酒厂仓库 Distillery Ware house
（照片由作者提供 Photo by author）

Bowmore is Islay's oldest distillery, and home to the world's oldest whisky maturation warehouse, the No.1 Vaults. It's in this legendary warehouse that Bowmore whisky has been meticulously matured by master distillers for nearly 240 years.

The first recorded mention of Bowmore Distillery dates from 1779, but speak kindly to a few of the older islanders and they'll maybe whisper a rumour or two that distilling started here some time before that.

Certainly Bowmore's founder, the merchant David Simpson, bought the land in 1766. Exactly what happened in the years in between is anyone's guess!

When it comes to whisky heritage, Bowmore Distillery is unrivalled. They continue to hand down the traditions, skills and expertise that were crafted in the earliest days of Scotch whisky.

Here are some important years for Bowmore Distillery:

1837. Glaswegian twin brothers William and James Mutter purchased Bowmore Distillery from the Simpson family. Their lasting legacy saw the building of the lade which draws water from the River Laggan.

1887年，在拥有了波摩蒸馏厂50年后，经济萧条使得威士忌需求量减少，穆特家族将蒸馏厂卖给了坎贝尔敦的约翰·贝尔·谢里夫。

1892年，蒸馏厂被约瑟夫·罗伯特·霍尔姆斯领导的财团收购，他们将蒸馏厂命名为波摩蒸馏厂公司。

1925年，詹姆斯·贝尔·谢里夫第二次收购了波摩蒸馏厂，并将其更名为谢里夫的波摩蒸馏厂。

1940年，波摩中断了生产，因为蒸馏厂被英国皇家空军海岸司令部征用，为第二次世界大战出力。1940年到1943年期间，这个蒸馏厂先后由三个中队控制。

1963年，威士忌经纪公司斯坦利·P.莫里森有限公司收购了波摩蒸馏厂。

1979年，为了庆祝蒸馏厂成立200周年，他们推出了一款稀少的限量版瓶装酒，反映出当时可用的原版穆特瓶。

1980年，英国女王伊丽莎白二世参观了波摩酒厂，这是她第一次参观苏格兰威士忌酒厂。她个人受赠一桶威士忌，这桶酒后来被装瓶卖掉，为当地的慈善机构筹集资金。

1993年，令人垂涎的"黑色波摩三部曲"的第一款上市了。

试饮和样品 Dram and samples（照片由作者提供 Photo by author）

1887. After just 50 years of owning Bowmore Distillery, economic depression saw a drop in demand for whisky, leading the Mutter family to sell the distillery to John Bell Sherriff of Campbeltown.

1892. The distillery was purchased by a consortium headed by Joseph Robert Holmes who named the distillery Bowmore Distillery Co..

1925. James Bell Sherriff bought Bowmore Distillery for the second time and renamed it Sherriff's Bowmore Distillery.

1940. Bowmore ceased production as the distillery was commandeered by RAF Costal Command to help with the efforts during the Second World War. Three squadrons operated the distillery between 1940 and 1943.

1963. Whisky broker Stanley P. Morrison Co. Ltd. purchased Bowmore Distillery.

1979. To celebrate the distillery's bicentenary, they released a rare and limited edition bottling that reflected the original Mutter bottles available at the time.

1980. Queen Elizabeth II visited Bowmore Distillery, her first visit to any scotch whisky distillery. She was presented with her own cask and bottles from this cask were later sold to raise money for local charities.

1993. The first expression of the coveted *Black Bowmore trilogy* was released.

2012年，1957年份酒款上市——这是所有上市的艾雷岛单一麦芽威士忌中最古老的。这款酒在一号仓库熟成了54年，仅有12瓶存世，使得这款威士忌成为极度珍稀的威士忌酒款。

2014年，宾三得利收购了莫里森波摩蒸馏者有限公司。

2015年，作为苏格兰单一麦芽威士忌的第一次，波摩水楢桶上市，它将苏格兰威士忌的力量和激情与日本水楢桶熟成的威士忌之优雅和精致完美地结合在了一起。

2016年，第五桶也是最后一桶黑色波摩被重新发现，结果推出了精妙的"黑色波摩50年，被重新发现的最后一桶"——这是对过去50年所有在酒厂工作过的匠人们的致敬。

酒厂全景 Panoramic view of the distillery（照片由作者提供 Photo by author）

2012. The 1957 expression was released—the oldest Islay single malt ever released. Matured in the No.1 Vaults for 54 years, only 12 bottles exist, making this an extremely rare whisky.

2014. Beam Suntory acquired Morrison Bowmore Distillers Ltd..

2015. In a first for single malt scotch whisky, Bowmore Mizunara was released, marrying the strength and passion of scotch whisky with the elegance and refinement of whisky matured in Japanese Mizunara casks.

2016. The fifth and final cask of Black Bowmore was rediscovered, resulting in the exquisite *Black Bowmore 50 Year Old, The Last Cask To Be Rediscovered*—a fitting tribute to all the craftsmen who have worked at the distillery over the past 50 years.

波摩一号 / Bowmore No.1

在第一次灌装的波本桶中熟成，释放出波摩一号的泥煤烟熏味与甜味。

40%ALC./vol.	
闻香	香草软糖、海洋空气以及泥煤烟熏味，被蜂巢和肉桂香料的气息完美地平衡了。
味觉	柑橘、温和的咸味以及香草和椰子片的味道。
余味	泥煤烟熏味、波本香草和酸橙味。

Bowmore

Maturing in first—fill Bourbon casks, unlock layers of peat smoke and sweetness in this No.1 malt.

40%ALC./vol.	
NOSE	Vanilla fudge, sea air and peat smoke, balanced beautifully by honeycomb and cinnamon spice.
TASTE	Citrus, gentle saltiness and vanilla with flakes of coconut.
FINISH	Peat smoke, bourbon vanilla and lime.

波摩12年 / Bowmore 12 Years Old

柠檬皮的气味让阵阵泥煤烟熏的气息以及一池池蜂蜜的气味，变得更香。

40%ALC./vol.

闻香	微妙的柠檬和蜂蜜味，与波摩标志性的泥煤烟熏味完美地平衡在一起。
味觉	温暖可口，带着些微黑巧克力的风味。
余味	标志性的柔和泥煤烟熏味，将人带入一个美味、悠长和圆润的尾调。

Bowmore

Puffs of peat smoke and pools of honey, sharpened by lemon zest.

40%ALC./vol.	
NOSE	Subtle lemon and honey, balanced beautifully by Bowmore's trademark peaty smokiness.
TASTE	Warm and delicious on the palate with subtle dark chocolate flavours.
FINISH	Trademark gentle peat smoke, leading to a delicious, long and mellow finish.

波摩15年 / Bowmore 15 Years Old

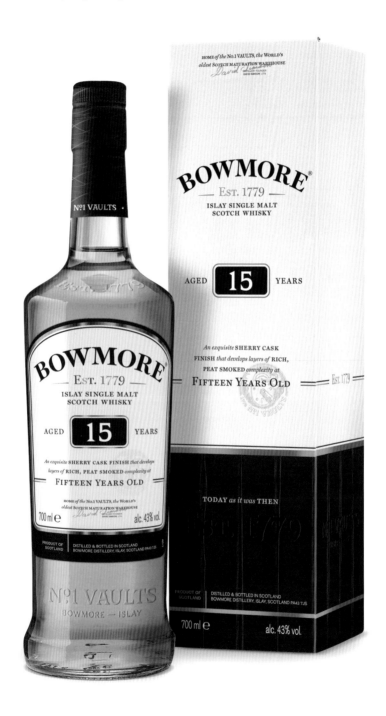

　　浓郁的葡萄干味以及柔和的烟熏味引领你到达美味的巧克力味中心。

43%ALC./vol.	
闻香	黑巧克力和葡萄干的香味与经典的波摩烟熏味相得益彰。
味觉	美妙的雪松木和浓郁的糖蜜太妃糖味道。
余味	辛辣，太妃糖、雪利酒和大麦味。

Bowmore

　　Rich raisins and gentle smoke lead the way to a delicious chocolaty centre.

43%ALC./vol.	
NOSE	Dark chocolate and raisin aromas compliment the classic Bowmore smokiness.
TASTE	Wonderful cedar wood and rich treacle toffee flavours.
FINISH	Spicy, toffee, sherry and barley.

波摩18年 / Bowmore 18 Years Old

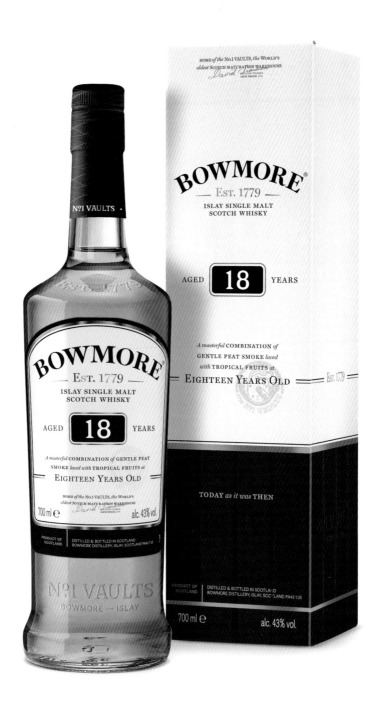

一个大师级的组合，包含了成熟水果、黑巧克力以及艾雷岛的烟熏味。

43%ALC./vol.	
闻香	奶油焦糖太妃糖味，带有成熟水果以及烟熏的芳香。
味觉	难以置信的复杂，口味是漂亮的浆果味以及巧克力味，与轻轻的烟熏味相平衡。
余味	持久而惊人的平衡之终章。

Bowmore

A masterful combination of ripe fruit, dark chocolate and Islay smoke.

43%ALC./vol.	
NOSE	Creamy caramel toffee, with ripe fruit and smoke aromas.
TASTE	Incredibly complex, with beautiful soft fruit and chocolate balanced with a light smokiness.
FINISH	The long and wonderfully balanced finish.

波摩10年旅行专款 /
Bowmore 10 Years Old Travel Exclusive

43%ALC./vol.

闻香	晒干的水果、烤过的塞维利亚橙和烤枫糖的气息。
味觉	大葡萄干、海盐以及深色水果油的味道。
余味	泥煤烟熏过的核桃和香料味道。

Bowmore

43%ALC./vol.	
NOSE	Sun-dried fruit, baked Seville orange and toasted maple.
TASTE	Sultanas, sea salt and dark fruit oils.
FINISH	Peat-smoked walnut and spices.

波摩15年旅行专款 /
Bowmore 15 Years Old Travel Exclusive

43%ALC./vol.	
闻香	西西里黄金甜点的气息，带着蜂蜜、阿玛菲柠檬和甜烟叶的气息。
味觉	丝绸般柔滑的柑橘、咸甜兼备的香草味以及蜂巢的味道。
余味	泥煤烟熏味、甜杏和绵长的柠檬味。

Bowmore

43%ALC./vol.	
NOSE	Sicilian golden desserts with honey, Amalfi lemon and sweet tobacco leaves.
TASTE	Silky citrus, salty-sweet vanilla and honeycomb.
FINISH	Peat smoke, sweet almond and lingering lemon.

波摩18年旅行专款 /
Bowmore 18 Years Old Travel Exclusive

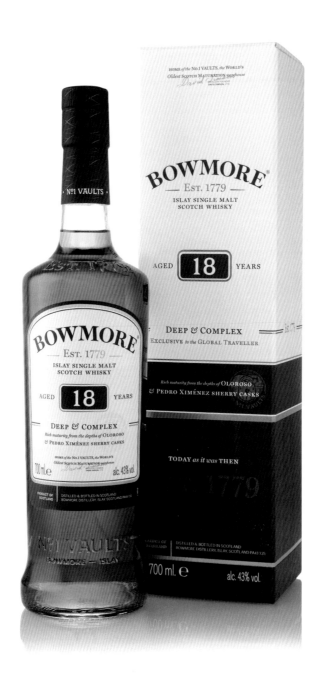

43%ALC./vol.

闻香	浓郁的黑巧克力、糖蜜太妃糖以及枣油的气息。
味觉	苦中带甜的橙子皮和泥煤熏烤咖啡的味道。
余味	天鹅绒摩卡和夏威夷坚果巧克力的韵味。

Bowmore

43%ALC./vol.	
NOSE	Rich dark chocolate, treacle toffee and date oil.
TASTE	Bitter-sweet orange peel and peat-smoke roasted coffee.
FINISH	Velvet mocha and macadamia nut chocolate.

布鲁赫拉迪 / **Bruichladdich**

　　布鲁赫拉迪酒厂是由哈维兄弟于1881年建造的，他们都是格拉斯哥威士忌家族中年轻有为的成员。

　　这个新的艾雷岛酒厂的志向，是为当时风靡的混合苏格兰威士忌提供非常重要的风味。布鲁赫拉迪酒厂当时被巧妙地设计成最先进的设施，但事实是，情况从一开始便背叛了它。哈维兄弟俩发生争吵，管理不善、禁酒令、大萧条和两次世界大战让它多次变更了所有权，却从未吸引到能让它在现代世界与同行竞争的投资。

蒸馏器 Stills
（照片由布鲁赫拉迪提供 Photo by Bruichladdich）

布鲁赫拉迪酒厂照片 Bruichladdich Distillery（照片由布鲁赫拉迪提供 Photo by Bruichladdich）

Bruichladdich Distillery was built in 1881 by the Harvey brothers who were young and ambitious members of Glasgow whisky family.

The purpose of the new Islay distillery was to provide the all-important flavour profiles for the blended Scotch whiskies that were becoming so popular at the time. Bruichladdich was cleverly designed as a state-of-the-art facility, but the truth was that circumstances conspired against it from the beginning. The brothers fell out, and poor management, prohibition, the depression and two World Wars saw it change ownership many times without ever attracting the investment that would have enabled it to compete in the modern world.

布鲁赫拉迪游客中心 Bruichladdich visitor centre（照片由作者提供 Photo by author）

　　关闭了一段时间后，这家当时已处于半废弃状态的酒厂于2000年被一群私人投资者收购，投资者们看到了这家老式酒厂的独特机遇。然后，村子里懂得手作工艺并懂得老式机械的人们自告奋勇而来。于是酒厂有了这样的决定：专注于手工制作小批量单一麦芽威士忌，利用人们的传统技能在组织内部做一切可能的事情。

　　随着威士忌产业的其他公司奉行产品标准化并压低雇佣率，布鲁赫拉迪挑战这些惯例，推出了一系列发人深省的高品质威士忌，蒸馏、熟成和装瓶均在艾雷岛上完成。

　　这家新公司借鉴了公司负责人对法国高档红酒和风土的精深了解，探索威士忌主要原料大麦的细微变化。艾雷岛的农民第一次被鼓励种植制酒用的大麦，来自不同农场和不同品种的谷物被分开发芽、分开蒸馏。这为威士忌世界引入一个全新的维度。后来，这家特立独行的新酒厂进一步激怒了威士忌的传统主义者，因为它推出了植物学家金酒，一种独特的艾雷岛干金酒。

After a period of closure the now semi-derelict distillery was purchased in 2000 by a group of private investors who saw a unique opportunity in the old fashioned plant. Men from the village then came forward who knew the artisanal methods and understood the old machinery. It was decided to concentrate on making small volumes of single malt by hand, doing everything possible in-house utilizing the traditional skills of the people.

As the rest of the whisky industry standardised their products and drove down employment, Bruichladdich challenged these conventions with a huge range of thought-provoking and very high quality whiskies that were distilled, matured and bottled on Islay.

Drawing on the principal's intimate knowledge of fine French wine and terroir, the new company explored the subtle variations in their main ingredient, barley. Islay farmers were encouraged to grow malting barley for the first time, with grain from different farms and different varieties being malted and distilled separately. This introduced a whole new dimension into the whisky world. The maverick newcomers then further infuriated whisky traditionalists by introducing the Botanist, a unique Islay dry gin.

酒厂鸟瞰景 Bird' s-eye view of the distillery（照片由布鲁赫拉迪提供 Photo by Bruichladdich）

　　威士忌行业的经济环境不适合懦夫，而这家自筹资金的独立公司总是挣扎着为其非凡的扩张筹集资金。法国高档烈酒巨头人头马君度集团在2012年向饱受煎熬的股东提出了一份无法拒绝的收购要约。令人高兴的是，新业主非常重视保持传统的生产方法，以及作为先进的赫布里底酒厂的布鲁赫拉迪酒厂所讲求的特立独行的精神。

The economics of the whisky industry are not for the faint hearted and the self-financed independent company always struggled to fund its extraordinary expansion. Its long-suffering shareholders were finally made an offer that was impossible to refuse by the French premium spirits giant Remy Cointreau in 2012. Happily the new owners place great emphasis on maintaining the traditional production methods and independently minded ethos of Bruichladdich as Progressive Hebridean Distillers.

酒厂入口 Distillery Entrance（照片由作者提供 Photo by author）

布鲁赫拉迪经典莱迪 /
Bruichladdich the Classic Laddie

50%ALC./vol.

闻香　以麦芽糖和少许薄荷为开端，引导我们进入最美妙的新鲜野花香气中：金凤花、雏菊、绣线菊、桃金娘、报春花和樱花。过了几秒钟，更多香气从杯子里升起，海浪飞沫的和风及海石竹的气息提醒你，这酒只可能是在海边熟成而来。四五分钟后，加入少许水，焦糖覆盖的水果气息就会浮出水面；呈现出柠檬汁混合蜂蜜、橘子和药片的气味。

味觉　甜橡木和大麦的味道混合在一起，使味蕾绽放。蒸馏的酒液蕴藏的水果味偶尔在大西洋的微风中闪现，并像香槟的泡泡一样在舌头上爆开。

余味　成熟的绿色水果、红糖和甜麦芽的组合迎来了结尾。

Bruichladdich

50%ALC./vol.
NOSE Opening with barley sugar and a hint of mint before leading into the most wonderful notes of freshly cut wild flowers: buttercup, daisy, meadowsweet, myrtle, primrose and cherry blossom. As the seconds tick by, more aromas rise from the glass, little zephyrs of spindrift and sea pinks reminding you that this spirit is matured exclusively by the sea. After some four or five minutes and with the addition of a little water, caramelised fruits drift onto the scene; lemon drops and honey, tangerine and tablet.
TASTE The sweet oak and the barley arriving together sending the taste buds into raptures. The fruits from distillation drift in on an Atlantic breeze and pop on the tongue like champagne bubbles.
FINISH A combination of ripe green fruit, brown sugar and sweet malt bring closure.

布鲁赫拉迪艾雷岛大麦2009 /
Bruichladdich Islay Barley 2009

50%ALC./vol.

闻香　布鲁赫拉迪酒液的花香特质立刻显现，它的纯净和开放让人耳目一新，振奋人心。一阵充满着精致花束花香的海风拂过，然后年轻、干净的布鲁赫拉迪的美，连同清脆的甜苹果、成熟的醋栗和桃子气息一同显现出来。大麦芽带来了红糖和太妃糖的香味，而来自烤橡木的香味成了平静而克制的背景，让麦芽味和酒液变得出众。奶油硬糖、香料和浓稠的香草奶油冻的气息。

味觉　口感的黏性，是一种享受。滴流蒸馏是一种缓慢而简单的技术，却是形成这种口感品质的基础。它的质地强健而丰富，天然油脂从谷物味中温和地协调了香料味，让它的年轻、干净、新鲜的性格魅力得以蓬勃发展。

余味　结尾是近海的清新空气，带着清爽的柠檬味和那股只有在大西洋沿岸熟成后才会带有的，如劲风吹来的盐味。

Bruichladdich

50%ALC./vol.	
NOSE	The floral nature of Bruichladdich spirit is immediately to the fore, its purity and openness at once refreshing and stimulating. A sea breeze filled with a delicate floral bouquet drifts through before the beauty of young, clean Bruichladdich emerges with crisp sweet apple, ripe gooseberry and peach. The malted barley brings brown sugar and toffee, while aromas from the toasted oak are calm and restrained as a backdrop, allowing the malt and spirit to shine. Butterscotch, spice and creamy vanilla custard.
TASTE	The mouthfeel, the viscosity, is a delight. Trickle distillation is a slow and simple technique but fundamental to producing such quality. The texture is muscular and rich, the natural oils coaxed gently from the grain temper the spice to allow its character, its young, clean, fresh charm to flourish.
FINISH	The finish is ozone fresh, with zesty lemon and that note of windswept salt that can only have come from maturation by Atlantic shoreline.

布鲁赫拉迪1984/32 /
Bruichladdich 1984/32

43.7%ALC./vol.

闻香	经典的老式布鲁赫拉迪。椰子、温暖的黑砂糖、香草奶油冻、葡萄柚、芒果、烤姜饼、杏仁软糖的气味。过了一会儿，就是糖浆和葡萄干的香气。
味觉	质地温和柔软，口感细腻。焦糖布丁、金凤花和橙子皮的味道，酒液轻柔地温暖着口腔，打开后呈现出一点皮革味及烤橡木和果仁的味道。甜柑橘和芒果的味道稍后出现。
余味	温暖的司康饼和干果、蜜橘、柠檬皮和杏桃果酱的余味是那么温柔而可爱。

Bruichladdich

43.7%ALC./vol.	
NOSE	Classic old school Bruichladdich. Coconut, warm muscovado sugar, vanilla custard, grapefruit, mango, baked ginger biscuits, marzipan. After a while notes of date syrup and sultana.
TASTE	Soft and gentle texture, a delicate dram. Crème brulee, buttercup and orange zest. Gently warming the palate, it opens to reveal a touch of leather, toasted oak and praline. Sweet citrus and mango come through later.
FINISH	So gentle but a lovely finish of warm scones and dried fruit, candied orange, lemon peel and apricot jam.

布鲁赫拉迪1985/32 /
Bruichladdich 1985/32

48.7%ALC./vol.

闻香	血橙、油桃、香蕉面包、椰蓉、香草奶油冻的香气，然后是更多的花香和坚果的气味，天竺葵、杏仁软糖、核桃、石南花蜜的香气。
味觉	在红酒桶中短暂的风味增强，赋予它一丝温暖的肉豆蔻的感觉，还有一层覆盆子、黑加仑果酱以及成熟李子的果香，所有元素都因在美国橡木桶中经年的时间而获得完美平衡——巧克力太妃糖、柠檬蛋白酥和椰蓉夹心巧克力棒的甜蜜。它不断地绽放，释放出一层又一层的复杂性。
余味	漫长而温柔，橡木气息强势影响着一波波柔和的香草威化、太妃糖和檀香木的芳香。然后是柠檬麦芽糖、少许咸味焦糖和杏桃糖浆的味道。

Bruichladdich

48.7%ALC./vol.	
NOSE	Blood orange, nectarines, banana bread, desiccated coconut, vanilla custard, then more floral and nutty tones, geranium, marzipan, walnut, heather honey.
TASTE	The short ACE in the wine casks has given a warm, nutmeg hint and a layer of fruity notes, raspberry, blackcurrant jam and ripe plum, all perfectly balanced by years in American oak— cinder toffee, lemon meringue and coconut— bounty bar sweet. It opens continually to reveal layer upon layer of complexity.
FINISH	Long and gentle, the oak influence is strong with waves of soft vanilla wafer, toffee and sandalwood. Then lemon barley sugars, a hint of salted caramel and apricot syrup.

布鲁赫拉迪1986/30 /
Bruichladdich 1986/30

44.6%ALC./vol.

闻香　　干果香——葡萄干、无花果糖浆、西梅和圣诞蛋糕的气息。黑砂糖、少许热砂和橘子蜜饯、干茶叶、烟斗丝和皮革的气味在PX雪利桶中糅合。

味觉　　在口感上进一步提高了体验，归功于PX雪利桶的甜味，葡萄干、无花果干、枣、葡萄糖的味道。皮革味、雪茄盒和圣诞节蛋糕的味道。所有在雪利桶中熟成的威士忌应有的它都有了，而且更多。它的复杂性和平衡性得以保留，因为它的海洋清新感和杏桃苹果的果香正来自布鲁赫拉迪的DNA中。

余味　　永无止境。甜甜的水果、PX雪利酒、盐焦糖、巧克力太妃糖和一丝烟味。

Bruichladdich

	44.6%ALC./vol.
NOSE	Dried fruit—raisins, fig syrup, prunes and Christmas cake. Muscovado sugar, a hint of hot sand and candied peel, dried tea leaves, pipe tobacco and leather rolled in PX.
TASTE	On the palate the experience raises the bar still further with sweet PX notes, raisins, dried figs, dates, grape sugar. Leather, cigar boxes and Christmas cake. Everything a sherry matured whisky should be and more. The complexity and balance is retained as the marine freshness and fruity notes of apricot and apple come through from the Bruichladdich DNA.
FINISH	Never-ending. Sweet fruit, PX, salted caramel, cinder toffee and a hint of smoke.

布鲁赫拉迪黑色艺术6.1 /
Bruichladdich Black Art 6.1

46.9%ALC./vol.

闻香　　初闻感受到巨大的深度和丰富性。用手指转动酒杯，唤醒它的芳香和浓郁的黑色焦橡木的气息，伴随着黑莓果酱、黑巧克力、葡萄干、李子、接骨木果和苹果的气息纷至沓来。杏仁蛋白酥和柠檬蛋白派给人甜蜜感，还诉说着布鲁赫拉迪的DNA与酒桶的合作无间。一开瓶，杉木和红糖的气息便跑了出来。

味觉　　橡木和水果气息充满丰富感和活力。酒体初绽就透出庄园果实的香气，与芳香的香草奶油冻的味道完美共处。再啜饮一口，品尝深色的果实；有大枣、无花果、葡萄干和巧克力、姜糖、红玫瑰的香味。无法被定义为任何一种特定的风格，这杯酒在不断地扭转和变化，每一层的释放更增加了"究竟这款威士忌是如何被创造出来"的神秘感。层叠的蜂窝、柔软的水果、果脯、烟草、椰子的味道。

余味　　巧克力、杏桃、菠萝和来自陈年的布鲁赫拉迪威士忌的经典异国水果韵诉说着橡木桶和原酒的品质和平衡感。橡木味用红糖、生姜坚果饼干、太妃糖、橙味焦糖和丝绒烟草的调子作为结尾，将其品质向我们娓娓道来。

Bruichladdich

46.9%ALC./vol.	
NOSE	There is a huge depth and richness on initial nosing. Roll the glass through your fingers to rouse the aromas and rich, black charred oak with blackberry jam, dark chocolate, raisin, plum, elderberry and apple appear. Notes of marzipan and lemon meringue pie give sweetness and tell of the Bruichladdich DNA working with the casks. Cedar wood and brown sugar come through as the whisky opens.
TASTE	The richness and vitality of the oak and the fruit. The soft orchard fruit of the spirit comes through after opening a little, sitting beautifully alongside fragrant vanilla custard. Sip again and taste the dark fruits; date, fig, raisin and chocolate, crystallised ginger and floral red rose. Impossible to define as a particular style this dram twists and changes constantly, each layer revealed adds to the mystery of how this whisky was created. Layers of honeycomb, soft fruit, praline, tobacco, coconut.
FINISH	Chocolate, apricot, pineapple, classic exotic fruits from well-aged Bruichladdich tell of the quality and balance of oak and spirit. The oak speaks of its quality now with brown sugar, ginger nut biscuits, toffee, orange scented caramel and a velvet tobacco finish.

夏洛特港10年 / Port Charlotte 10

46.9%ALC./vol.

闻香   烟熏味被海洋、新鲜空气的个性所缓和，时间赋予了橡木桶、烟熏味和原酒以平衡与和谐。烟熏味总是可被鼻子识别出来，它是干的，带着泥土味和泥煤灰风格的，所以它也容许橡木味伴随一波波金色焦糖、法奇软糖、香草奶油冻，少许姜、肉豆蔻和丁香的气息涌现出来。那里有柑橘类水果的气息，用一滴水即可从杯中引诱出柔和的柠檬蛋白霜和克莱门氏小柑橘的香气。深呼吸，野生百里香、石南花和海石竹的花香将你带到了大西洋的岸边。

味觉   在质地和风格上，是一种显而易见的细腻和柔软。再一次，风味平衡得极好，烟熏味将橡木深处带出的甜味松散地包裹着。椰子、香草奶油冻、柠檬蜂蜜混合着烟熏牡蛎以及被阳光炙烤的咸沙的味道。

余味   绝妙的结尾，烟熏味是必定的，但也有奶油软糖、麦芽、香橙、芒果和香蕉太妃派的柔和甜味，显示出橡木的深度和品质。每一口，多种层次在互换。随着烟熏味的发起与消逝，酒体的香气也生生灭灭，成熟苹果和杏桃香，漂亮地纠缠着麦芽和橡木的甜味，以及那种典型的夏洛特港的干烟熏味。

Bruichladdich

46.9%ALC./vol.	
NOSE	The smoke is calmed by the marine, ozone character, time has brought a balance. A harmony in the combination of oak, smoke and spirit. While the smoke is always discernible on the nose it is dry, earthy, peatash in its style and so allows the oak to come forward with waves of golden caramel, fudge, vanilla custard, hints of ginger, nutmeg and clove. There is citrus fruit, coaxed from the glass with a drop of water, gentle lemon meringue and clementine. Breathe deeply and the floral aromas of wild thyme, heather and sea pink bring you to this Atlantic coast.
TASTE	On the palate there is a noticeable delicacy and softness in texture and style. Again, the balance of flavour is superb as the smoke wraps loosely around the sweetness drawn from deep within the oak. Coconut, vanilla custard, lemon honey combines with smoked oysters and sun baked salty sand.
FINISH	The finish is sublime, smoky sure, but also the soft sweetness of fudge and malted barley, orange, mango and Banoffee pie hinting of the depth and quality of the oak. The many layers interchange on each sip. As the smoke comes and goes so too the notes of the spirit, ripe apple and apricot, beautifully intertwined with malt and oak sweetness and that typical Port Charlotte style dry smoke.

夏洛特港艾雷岛大麦2011 /
Port Charlotte Islay Barley 2011

50%ALC./vol.

闻香   充满活力的香气旋风从杯中释放，这是一种朴实的柏油泥煤烟味。柠檬、桃子和青葡萄香气从酒体中散发出来。香草荚、奶油椰子和巧克力气味诉说着在美国橡木桶中的熟成历史。香味的组合们纠缠推搡着，争着引起我们的注意。打开一段时间后，柠香蛋糕、白巧克力和蜜桃酸奶的香味会被烟熏味像天鹅绒毯一般包裹着升起。

味觉   少许盐味加强了海岸感。干燥烟雾、盐焦糖和苹果酱的味道，然后是柠檬汁、煮开的糖水和白胡椒粉的味道。烟熏味在口腔中是轻盈的，主要是焦油和木炭的调子，而非药剂的酚质感。接近尾声，这温和轻盈的酒体中展露出带着花香的海风，海洋气息沁人心脾。橡木和大麦的甜味与干燥泥煤的烟熏味非常相配。这些大麦在艾雷岛土生土长，品质似乎更加独特，并且带来了酒体中典型的蜜桃和苹果香。

余味   焦油夹杂着海水、泥煤烟雾和盐味、椰子和烟草味。闭上眼睛，这些烟雾将带你来到艾雷岛一个用浮木燃着篝火的沙滩上。

Bruichladdich

50%ALC./vol.	
NOSE	A whirlwind of vibrant aromas burst from the glass, an earthy tarred peat smoke. Lemon, peach and green grapes come from the spirit. Vanilla pods, creamed coconut and chocolate speak of the American oak maturation. The combinations of aromas intertwine and jostle for attention. some time after open allows notes of lemon drizzle cake, white chocolate and peach yoghurt to rise wrapped in a velvet blanket of light smoke.
TASTE	Hints of salt reinforce the coastal feel. Dry smoke, salted caramel and apple sauce. Then lemon drops, boiled sweets and ground white pepper. The smoke is light on the palate, tar and charcoal rather than medicinal phenols. Towards the finish and the light gentle spirit has an edge of floral sea breeze, marine and refreshing. The sweetness of the oak and the barley are matched beautifully by the dry peat smoke. Islay grown, the barley seems more distinct and brings through the peach and apple typical of our spirit.
FINISH	Tar and brine, peat smoke and salt, coconut and tobacco. Close your eyes and the smoke takes you to an Islay beach with a driftwood bonfire.

奥肯特摩09.1对白 / Octomore 09.1 Dialogos

奥肯特摩09.1对优质苏格兰单一麦芽威士忌的基本法则提出疑问。首席酿酒师展示了专注于原料配方以及深思熟虑的蒸馏是如何能够彻底打破"陈年越久就越好"的神话的。对本款酒的5年陈酿声明充满信心，这款09.1于2012年从100%苏格兰大麦中蒸馏而成，原酒全程在顶级的美国威士忌桶中熟成。

59.1%ALC./vol. 156 ppm

闻香	烟熏味和橡木气息，浓郁的甜味来自酒桶，有香草、巧克力太妃糖的气味。烟熏味很浓厚，如天鹅绒般柔软且带有植物香味、泥煤味和土质气味。
味觉	口感柔和——没有侵略性，坚果、软糖、牛轧糖的味道，细腻的椰子味和玫瑰花瓣的香味。
余味	长时间的泥煤余味，微微的糖浆、温暖的沙子和盐的韵味。

Bruichladdich

Octomore 09.1 questions the fundamentals of superior single malt Scotch. Confident in its five-year-old statement, Head Distiller demonstrates how a focus on raw ingredients and considered distillation can definitively debunk the myth that older is better. Distilled in 2012 from 100% Scottish barley, the 09.1 is matured full-term in top-tier ex-American whiskey casks.

59.1%ALC./vol. 156 ppm	
NOSE	Smoke and oak, intense sweetness from the cask, vanilla, cinder toffee. The smoke is thick, velvety and vegetal, peat and earthy.
TASTE	Soft on the palate—no aggression, nutty, fudge, nougat, delicate coconut and rose petal floral.
FINISH	Long peat finish, hints of treacle, warm sand and salt.

奥肯特摩09.2对白 / Octomore 09.2 Dialogos

奥肯特摩09.2在其最后一年之前一直遵循与09.1版本相同的路径。蒸馏于2012年，这款强劲的液体花了4年时间在美国橡木桶中熟成，直到被重新灌入最好的法国橡木桶中。这些橡木桶是从波尔多地区精心挑选而来，让干燥泥煤烟熏味与甜水果的风味结合成一个专家级的组合。

58.2%ALC./vol. 156 ppm

闻香	烟熏味的果酱，黑加仑、樱桃、杏仁的气息。烟熏味带着土质感，咖啡渣、橡木和香草气息全部混合在一起并且总是有草莓和黑加仑果酱的香气。
味觉	黑暗且复杂，黑樱桃、土质烟熏味、苦咖啡、成熟李子和烟丝的味道。
余味	香料味贯穿，土质感、辛辣感、烟熏味一如既往的强烈。

Bruichladdich

Octomore 09.2 has followed the same path as its 09.1 counterpart until its final year. Distilled in 2012, this muscular liquid has spent four years maturing in ex-American oak casks before being recasked into the finest French oak. Carefully selected from the Bordeaux region, these casks coax the dry peat smoke into an expert combination with sweet fruit flavours.

58.2%ALC./vol. 156 ppm	
NOSE	Smoked jam, blackcurrant, cherry, almond. The smoke is earthy, coffee grounds, oak and vanilla all mixed and there is always fruity strawberry and blackcurranty jam.
TASTE	Dark and complex on the palate, dark cherry, earthy smoke, bitter coffee. Ripe plums and pipe tobacco.
FINISH	Spice comes through, earthy, peppery, smoke as ever comes through strong on the palate.

奥肯特摩09.3对白 / Octomore 09.3 Dialogos

仅52吨珍贵的艾雷岛大麦被制成麦芽，用于生产这款单一农场的奥肯特摩09.3版本威士忌。这些大麦2011年在奥肯特摩农场的艾琳的田里长大，这种来自超级原产地的单一麦芽后来被蒸馏成酒液并装入134个桶中。主要因为二次灌装，木桶产生的柔和影响，容许了更多的本地生大麦风味切入传说中的泥煤烟熏味中。

62.9%ALC./vol. 133 ppm

闻香	麦芽、好立克热饮，麦芽浆进入麦芽浆桶的气味。红茶、金色糖浆、柠檬海绵蛋糕的香气。
味觉	罐装金色糖浆，烟熏的茶味，麦芽汁、麦芽、梨、蜜桃冰激凌、海雾、石南花的味道。
余味	烟熏味、烤橡木、黑麦面包、盐、臭氧味。

Bruichladdich

Just 52 tonnes of precious Islay barley were malted to produce this single farm Octomore 09.3 edition. Raised in Irene's Field on Octomore Farm in 2011, this uber-provenance single malt was later distilled and filled into 134 casks. Predominantly second-fill, the gentler influence of the wood allows more of the locally grown barley flavours to cut through the legendary peat smoke.

62.9%ALC./vol. 133 ppm	
NOSE	Malted barley, Horlicks, the smell of the mash going into the mash tun. Black tea, golden syrup, lemon sponge.
TASTE	Golden syrup in the tin, smoked tea, wort, malted barley, pear, peach melba. Sea spray, heather flowers.
FINISH	Smoke, toasted oak, rye bread, salt, ozone.

布纳哈本 / **Bunnahabhain**

岛屿景观 Islay landscape
（照片由作者提供 Photo by author）

  布纳哈本——在盖尔语中是"河口"的意思，因为它位于艾雷海峡海岸上的玛加岱尔泉边，建立于1883年。然而，它的起源可以追溯到1879年，来自罗伯逊和百特调和屋的威廉·罗伯逊联合格林利斯兄弟创建了艾雷酒厂。所以，1883年，随着靠近玛加岱尔河的酒厂的建成，布纳哈本就诞生了。

  早年，酒厂依靠海上贸易。他们仅靠一个小村庄、一个码头和许多懂得如何酿造威士忌的人来支撑自己，用船接收补给，并派出冒险的海员，带着装满"好东西"的木桶沿着艾雷海峡销回大陆。

  1930年，由于在第一次世界大战后，整个欧洲仍在经济低迷中挣扎，布纳哈本酒厂就关闭了。然而，多亏了布纳哈本精神，仅7年后，它又重新开启。

  尽管艾雷岛的大部分地区没变，但在1960年，他们经历的最大变化发生了，一条新路建成了！沿艾雷海峡行驶的河豚船长久以来一直是生命线，但随着公路的引入，现在可以通过陆路运送补给。

  随着布纳哈本的消息传开，人们对威士忌的需求猛增，1963年，酿酒厂又安装了第二对蒸馏器，以扩大生产能力。

Bunnahabhain—which means "mouth of the river" in Gaelic, as it stands at the mouth of the Margadale Spring on the shores of the Sound of Islay—first came into existence in 1883. However, its origins can be traced back four years earlier when, in 1879, William Robertson of Robertson and Baxter Blending House, joined with the Greenlees Brothers to create the Islay Distillery Company. And so, with the distillery built on a site close to the Margadale River, in 1883 Bunnahabhain was born.

In the early years, distillery relied upon the sea trade. Armed only with a small village, a pier and lots of whisky making know-how, they received supplies by boat and sent adventuring seafarers back to the mainland along the Sound of Islay with casks of the "good stuff".

In 1930, with the whole of Europe still reeling from the economic downturn, in the aftermath of the Great War, Bunnahabhain Distillery closed its doors. However, thanks to the Bunnahabhain spirit, it was opened once again, a mere 7 years later.

While much of Islay hasn't changed, in 1960 one of the biggest changes they ever experienced occurred: a new road was built! The puffer boats travelling along the Sound of Islay had been the lifeline for so long, but with the introduction of the road, supplies could now reach by land.

As word of Bunnahabhain spread and demand soared, a second pair of stills were installed in 1963 to increase the production capacity.

1979年，经典12年陈酿苏格兰纯麦威士忌问世，受到了广泛的好评。

1993年，最后一艘船靠岸，标志着布纳哈本的一个历史性时刻。通过河豚船接收补给一个多世纪之后，新修的这条路被认为是最适合的路线，用来接收不断增加的原料和补给。

2006年，在被邦史都华蒸馏者有限公司收购3年后，该公司推出了一次重大重塑，重新设计了12年的包装，并推出了18年和25年的改型。

2010年是布纳哈本进化的又一个历史性时刻。它的威士忌又恢复了非冷滤的生产方式，色泽自然，酒精度在46.3%——就像过去一样。

2014年，邦史都华与迪思特合并，为该品牌和酒厂带来了更多投资。2017年，产品组合进一步扩大，增加了两个变种：舵手和双重烟雾。

岛屿景观 Islay landscape（照片由作者提供 Photo by author）

In 1979, classic 12 Years Old Single Malt Scotch Whisky was introduced to the world, received to great acclaim.

1993 marked a historic moment in Bunnahabhain's history as the last boat docked. After receiving supplies via puffer boats travelling along the Sound of Islay for over a century, the road was deemed the most suitable route for receiving ever increasing ingredients and supplies.

In 2006, three years after being purchased by Burn Stewart Distillers, a major rebrand was launched, with a redesign of the 12 Year Old packaging, and the launch of 18 Years Old and 25 Years Old variants.

2010 marked yet another historic moment in the evolution of Bunnahabhain. The whiskies returned to being produced non-chill filtered, with a natural colour at 46.3%— just as they had been in old time.

In 2014, Burn Stewart merged with Distell, which heralded more investment into both the brand and the distillery. In 2017, the portfolio of products was extended with two further variants: Stiùireadair & Toiteach A Dhà.

布纳哈本舵手 /
Bunnahabhain Stiùireadair

46.3%ALC./vol.	
闻香	干果和奶油焦糖，略带海水味、香草味、坚果味和少许香料味。
味觉	口感柔滑，有干果味，海盐、奶油焦糖带一丝坚果以及温和的香料味。
余味	回味悠长，有一丝干果的味道。

Bunnahabhain

46.3%ALC./vol.	
NOSE	Dried fruit and creamy caramel with hints of brine, vanilla, nuts and a touch of spice.
TASTE	Creamy mouth feel with a dried fruit influence, sea salt, creamy caramel with hints of nuts and a gentle spice.
FINISH	Long and lingering with hints of dried fruit.

布纳哈本双重烟雾 /
Bunnahabhain Toiteach A Dhà

46.3%ALC./vol.

闻香	迷人的泥煤味，略带雪利酒和浓郁的橡木味。
味觉	瞬间变暖的泥煤味转向甜美的欧罗索雪利酒的味道，混合了浓郁的橡木味，并被微妙的胡椒味所平衡。
余味	强健的尾韵长度经得起挑剔的品鉴。

Bunnahabhain

46.3%ALC./vol.	
NOSE	Intriguing peatiness with hints of sherry and rich oak.
TASTE	Immediate warming peatiness drift to a sweet Oloroso sherry influence with rich oak, balanced with delicate pepper.
FINISH	Robust length for the discerning palate.

布纳哈本12年 /
Bunnahabhain 12 Years Old

46.3%ALC./vol.

闻香	清新芳香，果味花香，带有一丝干果味和淡淡的烟味。
味觉	轻盈，带果香，坚果风味带着一丝香草和焦糖的香甜。
余味	余韵悠长，酒体丰腴。

Bunnahabhain

46.3%ALC./vol.	
NOSE	Fresh and aromatic, fruity floral with hints of dried fruit and a subtle prevalence of smoke.
TASTE	Light with fruit notes, nutty flavours with a sweetness and slight hints of vanilla and caramel.
FINISH	Lingering, beautifully rich and full-bodied.

布纳哈本18年 /
Bunnahabhain 18 Years Old

46.3%ALC./vol.	
闻香	散发着干果、蜜渍坚果、太妃糖和少许香料的芬芳。
味觉	芳醇的雪利酒果仁、干果和浓郁的橡木味，带有太妃糖和些微来自于在海边熟成18年而带有的淡淡盐味。
余味	干果味，略带香料和盐味。

Bunnahabhain

46.3%ALC./vol.	
NOSE	Fragrant with notes of dried fruit, rich honeyed nuts, toffee and slight hints of spice.
TASTE	Notes of mellow sherried nuts, dried fruit and rich oak, with toffee and a slight tang of salt from its maturation of 18 years at the coast.
FINISH	A dry fruit finish with a hint of spice and salt.

布纳哈本25年 /
Bunnahabhain 25 Years Old

46.3%ALC./vol.	
闻香	有浓郁的雪利酒香气，散发出香甜的焦糖和抛光皮革的香味。
味觉	最初是甜浆果和奶油味，后来演变成一顿烤坚果和麦芽的盛宴。
余味	柔和干爽，带着一丝微妙的甜味以及含香料的橡木味，回味悠长。

Bunnahabhain

46.3%ALC./vol.	
NOSE	Scents of rich sherry, blossoming into sweet caramel and polished leather.
TASTE	An initial taste of sweet berries and cream that evolves into a feast of roasted nuts and malts.
FINISH	Soft and dry with a delicate hint of sweet sugar and spiced oak that linger long on the tongue.

布纳哈本40年 /
Bunnahabhain 40 Years Old

41.9%ALC./vol.

闻香	带有热带水果、香蕉、浆果、奶油太妃糖、香草、浓郁坚果味和微甜橡木味。
味觉	带有甜麦芽、奶油焦糖、香草、奶油浆果、烤坚果、热带水果、香蕉和菠萝的味道。
余味	悠长的甜味和果香。

Bunnahabhain

41.9%ALC./vol.	
NOSE	Hints of tropical fruits, banana, berries, creamy toffee, vanilla, rich nuttiness and delicate sweet oak.
TASTE	Sweet malt, creamy caramel, vanilla, creamy berries, with hints of toasted nuts, tropical fruit, banana and pineapple.
FINISH	Long sweet and fruity.

布纳哈本1997年帕罗科塔多桶陈 /
Bunnahabhain 1997 Palo Cortado Cask Finish

54.9%ALC./vol.

闻香	新鲜的浆果和奶油、胡桃和核桃味，浓郁的橡木味和香草味，淡淡的干果味。
味觉	甜干果、浓郁橡木、烤坚果、淡淡的可可味，覆盖口腔。
余味	稍干的尾韵，略带肉豆蔻味。

Bunnahabhain

54.9%ALC./vol.	
NOSE	Fresh berries and cream, pecan and walnuts, rich oak with vanilla, hints of dried fruit.
TASTE	Mouth coating, sweet dried fruit, rich oak, roasted nuts, subtle hints of cocoa.
FINISH	Slightly dry finish with hints of nutmeg.

卡尔里拉 / Caol Ila

　　卡尔里拉由格拉斯哥酒厂厂主赫克托·亨德森于1846年创建，名为盖尔语"Caol Ila"，意为"艾雷海峡"。1879年，蒸馏厂重建，合并了自己的码头，蒸汽船可以在那里卸下供给并装载威士忌销往大陆。尽管在历史上经历过几次关闭和限制，并且艰难地增加产量，卡尔里拉还是在1974年全面重建，以满足混合威士忌持续增长的需求，使其成为艾雷岛上最大的蒸馏厂。

酒厂照片 Distillery（照片由作者提供 Photo by author）

Caol Ila was founded by Glasgow Distillery owner Hector Henderson in 1846, named after the Gaelic Caol Ila which means "Sound of Islay". By 1879, the distillery is rebuilt incorporating its own pier where steamships can unload supplies and load up on whisky for sale on the mainland. Despite suffering a few closures and restrictions throughout its history and struggling to increase production, Caol Ila was completely rebuilt in 1974 to meet the increasing blended whisky demand, making it the biggest distillery on Islay.

酒厂码头景色View of distillery pier
（照片由作者提供 Photo by author）

卡尔里拉的一些重要年份

1863年——贸易商布洛克·拉得公司接手。此时混合威士忌市场火爆，生意很好。

1879年——在超过20年的时间里，经历3次易手，见证了蒸馏厂重建并扩大。到1879年的时候卡尔里拉拥有了自己的码头，可以让蒸汽船和河豚船卸下补给货物并装载威士忌销往大陆。

1920年——布洛克·拉得作为一家公司进入清算程序，被卖给了J.P奥伯里恩有限公司，这个公司又将其卖给了一个财团——卡尔里拉蒸馏厂有限公司。

1927年——蒸馏者有限公司取得了卡尔里拉的控股权。

1934年——酒厂重开，重新开始生产。

1941年——对人力和大麦的限制意味着工厂不得不关闭。

1972年——随着酒厂从两个蒸馏炉扩大到六个蒸馏炉，产量也随之增加。

1974年——市场对混合威士忌的需求再次影响了卡尔里拉的命运，为了满足不断增长的需求，该公司进行了彻底的重建——让它成了艾雷岛最大的酒厂。

Some important years for Caol Ila Distillery

1863—Traders Bulloch Lade & Co. take over. The market for blended whisky is booming and business is good.

1879—Over two decades, three changes of hands sees the distillery rebuilt and expanded. By 1879 Caol Ila has its own pier where steam ships or puffers can unload supplies and load up on whisky for sale on the mainland.

1920—Bulloch Lade goes into liquidation as a company and is sold to J.P. O'Brien Ltd., who sells it to a consortium—Caol Ila Distillery Co. Ltd..

1927—The Distillers Company Ltd. lands the controlling interest in Caol Ila.

1934—The distillery is reopened and production begins again.

1941—Restrictions on people power and barley mean the distillery has to close.

1972—Production increases as the distillery expands from two to six stills.

1974—A demand for blended whisky once again influences the fate of Caol Ila, which is completely rebuilt to meet increasing demand—making it the biggest on Islay.

卡尔里拉黎明 / Caol Ila Moch

一支绝对直截了当的酒，卡尔里拉不说废话。爽口，果断，直接且明白无误。

43%ALC./vol.	
闻香	像是退潮后潮湿、带咸味的沙子，有消毒药水味。随后有一丝烟熏味，就像来自一堆古老的篝火那样。
味觉	未经淡化，甜味和咸味带着适度的酸味，卡尔里拉标志性的烟熏调的消毒水味就在这里。
余味	很短，干、温暖中带点苦味。萦绕着烟熏味的回味。

Caol Ila

An absolutely straightforward, no-nonsense Caol Ila. Crisp, assertive, direct and unmistakable.

43%ALC./vol.	
NOSE	Like wet, salty sand after the tide has receded, with antiseptic notes. Some smoke behind, as from an old bonfire.
TASTE	Unreduced, sweet and salty with a balancing acidity, Caol Ila's trademark smoky-antiseptic character is here.
FINISH	Quite short, drying and warming with a little bitterness. Lingering smoky aftertaste.

卡尔里拉12年 / Caol Ila 12 Years Old

卡尔里拉12年单一麦芽苏格兰威士忌呈淡稻草色，口感细腻平衡。

43%ALC./vol.

闻香	柔和的柠檬果香；一点点沐浴油和漱口水的气味。清新开胃，几乎没有烟熏味。
味觉	在自然的强度下口感良好；开头甜蜜；令人愉悦、轻盈芳香的烟熏味回味悠长。
余味	香甜的烟熏味萦绕，微酸的结尾。

Caol Ila

Caol Ila 12 Years Old Single Malt Scotch Whisky is the colour of pale straw, with a delicate balance of tastes.

43%ALC./vol.	
NOSE	Subdued, citric fruitiness; a whiff of bath oil and dentist's mouthwash. A fresh and appetising nose, with little or no trace of smoke.
TASTE	Drinks well at natural strength; sweet start; pleasant, light fragrant smokiness and a lengthy finish.
FINISH	Sweet smokiness in the lingering, slightly sour finish.

卡尔里拉酒厂限量款 /
Caol Ila Distillers Edition

丰富的风味，然而味道不甜，平衡得很好，莫斯卡托桶的痕迹在这款双重熟成的卡尔里拉酒款中不会过于明显。

43%ALC./vol.	
闻香	漂亮地集中并且干净利落，泥煤味、药水味，带着丰富的果香，辛辣而芳香。
味觉	甜麦芽味首先冲出来，接着被泥煤烟熏味淹没，具有强烈的、爽口的风味，还有肉桂香料味。
余味	绵长、圆润、醇厚、多层次。

Caol Ila

Richly flavoured yet also drying and finely balanced, the Moscatel cask wood is not over evident in this double-matured Caol Ila.

43%ALC./vol.	
NOSE	Wonderfully concentrated and clean, peaty, medicinal, with rich fruit, spicy and fragrant.
TASTE	Sweet maltiness strike first, then overwhelmed by peat smoke, intense, crisp flavours, and cinnamon spice.
FINISH	Long, rounded, robust and multi-layered.

卡尔里拉18年 / Caol Ila 18 Years Old

自鼻腔开始是烟熏调的甜蜜气息，接着口感平顺，表现出一种酸甜相间的特点。持久的结尾唤起一个遥远的、烟雾缭绕的海边篝火的印象。

43%ALC./vol.

闻香	烟熏的篝火气息，然后是肥皂水和湿羊毛，还有远处海滩上冒着烟的篝火。略带矿物油的味道，然后是蜡。
味觉	柔顺温和的甜味打头，带着很好的酸度和海盐味道。
余味	余韵悠长，烧焦的、略带酸味的收尾。

Caol Ila

Starts smoky-sweet on the nose then drinks smoothly, showing a sweet yet sour character. The long-lived finish evokes a distant, smouldering beach bonfire.

43%ALC./vol.	
NOSE	Smoky bonfires, then soapy water and wet wool, with a smouldering beach bonfire in the distance. Hints of mineral oil, then wax.
TASTE	Smooth and mildly sweet start with good acidity and sea salt.
FINISH	A lingering, charred, slightly sour finish.

生命之水：艾雷岛威士忌品鉴指南；汉英对照

....
The Water of Life : A Tasting Guide to **Islay Whisky** : Chinese-English Bilingual Version

卡尔里拉桶强 /
Caol Ila Cask Strength

口感紧实顺滑，带着甜味，又转化为海盐味。一种令人愉悦的、持久的、微焦的收尾。通常会将原桶强度加水稀释，但是不要加太多，因为这样会变得过淡。

变化的%ALC./vol.	
闻香	消毒水味，随后是海边篝火的烟熏味，接着是带药水味的淡淡花香。
味觉	被淡化，柔顺且温和的甜味；轻微酸味和咸味。与自然强度相比，温和。
余味	令人愉悦、微焦、延绵收尾。

Caol Ila

Firm and smooth on the palate with sweetness again turning to sea saltiness. A pleasant, sustained, lightly charred finish. It's usual with cask strength malts to add water, but don't add too much as it can be surprisingly mellow.

VARIES%ALC./vol.	
NOSE	Carbolic, with a beach-bonfire smokiness behind. Then medicinal, lightly floral.
TASTE	Reduced, smooth and mildly sweet; light acidity and salty. Mild, compared to the flavour at natural strength.
FINISH	Pleasant, lightly charred, lingering.

卡尔里拉25年 /
Caol Ila 25 Years Old

一款为漫长之夜而准备的
酒。一款平衡性良好、经过深
思熟虑的老式卡尔里拉；篝火
还在燃烧，它的烈度因年岁而
降低。

43%ALC./vol.

闻香	甜香料和清新海风的香气丝毫不减。也许还有一丝果脯气息。加一点水，会带出一种微妙的、芳香的抛光后的木材气息。
味觉	毫不褪色，迎面而来的甜味和水果味，带着一种平衡的苦味。与它的年份相适应的是，卡尔里拉标志性的烟熏—药水味得以控制，而不是充溢出来。口感良好，辅以复杂的风味。
余味	很长，甜调的干味，更多的烟熏味传出。令人愉悦的回味。

Caol Ila

One for the long evenings. A well-balanced, contemplative old Caol Ila; the bonfire still burns, its intensity dimmed by age.

43%ALC./vol.	
NOSE	Unreduced, sweet-spicy and fresh sea-air notes. Hints of preserved fruit, perhaps. Adding a little water brings up subtle, fragrant polished wood.
TASTE	Unreduced, immediately sweet and fruity with a balancing bitterness. As befits its age, Caol Ila's trademark smoky-medicinal character is controlled rather than exuberant. A good mouthfeel complemented by a complex flavour.
FINISH	Quite long, sweetly drying, with more smoke coming through. Pleasant aftertaste.

齐侯门 / **Kilchoman**

大麦田 Barley field（照片由齐侯门提供 Photo by Kilchoman）

齐侯门蒸馏厂是一家小型家庭经营的农场式蒸馏厂，坐落于艾雷岛西北海岸克雷格摩尔岬庇护下的岩边农场的建筑之间。得益于艾雷岛的农场式蒸馏厂的出名，齐侯门的大麦田向西延伸至马希尔湾和大西洋海岸。

这家酒厂由安东尼·威尔斯在2005年建立，他的想法是建造一座新酒厂，让威士忌酿造回归本源。

自那以后，安东尼和妻子凯西之间又加入了三个成员：儿子乔治、詹姆斯和彼得。齐侯门以种植在酒厂周围土地里的大麦为原料，生产100% 艾雷岛单一麦芽威士忌，并且，威士忌酿造的每个阶段都在酒厂内完成，包括传统的地板麦芽制造法工序。

Kilchoman Distillery is a small family run farm distillery nestled amongst the farm buildings of Rockside Farm sheltered beneath the Creag Mhor headland on the north-west coast of Islay. Famous as Islay's Farm Distillery, the Kilchoman barley fields stretch west to the shores of Machir Bay and the Atlantic Ocean.

The distillery was established by Anthony Wills in 2005. His idea was to build a new distillery which took whisky production back to its roots.

Since then Anthony and wife Kathy have been joined by their three sons: George, James and Peter. Kilchoman produces its 100% Islay single malt from barley which is grown in fields surrounding the distillery and completes every stage of the whisky-making process at the distillery, including traditional floor malting.

酒厂鸟瞰景 Distillery bird's-eye view（照片由齐侯门提供 Photo by Kilchoman）

　　齐侯门是苏格兰唯一使用此法的酒厂。自2009年第一次装瓶以来，齐侯门迅速成为艾雷岛高品质单一麦芽威士忌品牌，并获得以下这些显著奖项：

　　——在久负盛名的国际威士忌大赛（IWC）上，安东尼·威尔斯被评为2013年度最佳蒸馏大师，齐侯门的卓越品质得到认可。

　　——2016年，两个核心酒款都赢得了主要奖项，塞纳滩获得了国际葡萄酒与烈酒大赛的最佳无年份苏格兰单一纯麦威士忌奖项，而马希尔湾赢得了国际威士忌大赛的最佳艾雷岛单一纯麦威士忌奖项。

　　——2017年，酒厂的旗舰酒款——马希尔湾，被评为威士忌交易所2017年度威士忌，同时，一款在朗姆桶内熟成的威士忌赢得了吉姆莫瑞的《威士忌圣经》2017年度单桶威士忌奖项。

产品 Priducts（照片由齐侯门提供 Photo by Kilchoman）

从酒桶中提取的试饮 Dram from the barrel
（照片由齐侯门提供 Photo by Kilchoman）

Kilchoman is the only distillery in Scotland to do this. Since the first bottling was released in 2009, Kilchoman has quickly established itself as a high-quality Islay Single Malt Whisky with these notable awards:

—Kilchoman's exceptional quality was recognised at the prestigious International Whisky Competition (IWC) when Anthony Wills was named Master Distiller of the Year 2013.

—In 2016, both core expressions won major awards with Sanaig named Best Single Malt Scotch Whisky—no age statement at the IWSC and Machir Bay came awarded Best Islay Single Malt Scotch in the International Whisky Competition.

—In 2017 the distillery's flagship expression, Machir Bay, was named Whisky Exchange Whisky of the Year 2017. Also, a rum matured single cask won Jim Murray's *Whisky Bible*, Single Cask of the Year 2017.

仓库 Warehouse（照片由作者提供 Photo by author）

齐侯门马希尔湾 / Kilchoman Machir Bay

马希尔湾是齐侯门产品线的旗舰款，它是波本桶和雪利桶熟成的齐侯门单一麦芽威士忌的结合。大约80%来自波本桶，20%来自欧罗索雪利桶。

酒款马希尔湾以艾雷岛上最壮观的海滩而得名。

46%ALC./vol.	
闻香	柔和的烹熟果香以及强劲的泥煤味。
味觉	混合果香、香草味和强烈的甜味。
余味	回味绵长。

Machir Bay is the flagship expression of the Kilchoman range. It is a combination of both bourbon and sherry cask matured Kilchoman single malt. Approximately 80% bourbon barrels and 20% Oloroso sherry casks.

Machir Bay is named after the most spectacular beach on Islay.

46%ALC./vol.	
NOSE	Soft cooked fruits with strong peaty aromas.
TASTE	Mixed fruits and vanilla with intense sweetness.
FINISH	A long lingering finish.

齐侯门塞纳滩 / Kilchoman Sanaig

塞纳滩是主要由雪利桶影响的核心酒款，70%来自欧罗索雪利桶，30%来自波本桶。

酒款塞纳滩以酒厂北部一个遍布岩石的小水湾而得名。

46%ALC./vol.	
闻香	柔和的烹熟果香以及焦糖和香草气味。
味觉	太妃糖、泥煤烟熏和柑橘香，绵长的甜味。
余味	泥煤烟熏、果香和甜味的完美平衡。

Sanaig is predominantly sherry cask influenced core expression, a vatting of 70% oloroso sherry casks and 30% bourbon barrels.

Sanaig is named after the rocky inlet just north of the distillery.

46%ALC./vol.	
NOSE	Soft cooked fruits with caramel and vanilla.
TASTE	Toffee, peat smoke and citrus with lingering sweetness.
FINISH	A lovely balance of peat smoke, fruit and sweetness.

齐侯门格姆湖2018版 /
Kilchoman Loch Gorm 2018 Edition

酒桶类型：欧罗索雪利桶

泥煤含量等级： 50 ppm

限量版： 15,000瓶

格姆湖2018版取自19个分别在2007年、2008年、2011年装桶的欧罗索雪利桶。

酒款格姆湖得名于齐侯门酒厂北部的湖，此湖色深，呈泥炭色，类似于熟成在齐侯门雪利桶中的单一麦芽威士忌的深邃丰富之色。

43%ALC./vol.

闻香	橙皮、丁香、混合香料以及烹熟果香。
味觉	丰富香辛味、烹熟果香和泥煤烟熏味的漂亮平衡。
余味	口腔充满泥煤烟熏味、持久的热带甜味以及丰富的干果韵味。

CASK TYPE: OLOROSO SHERRY BUTTS
PEATING LEVEL: 50 ppm
LIMITED EDITION: 15,000 BOTTLES

The Loch Gorm 2018 Edition is a vatting of 19 Oloroso sherry butts filled in 2007, 2008 and 2011.

Loch Gorm is named after the loch located to the north of the distillery, very dark and peaty in colour, similar to the dark rich colour of Kilchoman's sherry cask matured single malt.

43%ALC./vol.	
NOSE	Orange peel, cloves, mixed spice and cooked fruits.
TASTE	A beautiful balance of spicy richness, cooked fruits and peat smoke.
FINISH	Mouth filling peat smoke, lasting tropical sweetness and rich dried fruit.

齐侯门波特桶熟成2018版 /
Kilchoman Port Cask Matured 2018 Edition

酒桶类型：红宝石波特桶
泥煤含量等级：50 ppm
限量版：10,000瓶

2018版波特桶熟成酒款来自30桶2014年装在红宝石波特桶中熟成的威士忌。

50%ALC./vol.

闻香	红色水果，带有淡淡的柑橘香，泥煤味，以及淡淡的奶油甜香。
味觉	柑橘、花香的甜味和海岸的层层味道中，带有红加仑果酱和肉桂的味道。
余味	红色水果经烹熟后的层层韵味。

CASK TYPE: RUBY PORT HOGSHEADS
PEATING LEVEL: 50 ppm
LIMITED EDITION: 10,000 BOTTLES

The Port Cask Matured 2018 Edition is a vatting of 30 ruby port hogsheads filled in 2014.

50%ALC./vol.	
NOSE	Red fruits with light citrus, earthy peat smoke and hints of creamy sweetness.
TASTE	Redcurrant jam, cinnamon with layers of citrus, floral sweetness and coastal influence.
FINISH	Layers of cooked red fruits.

齐侯门100%艾雷第8版 /
Kilchoman 100% Islay 8th Edition

酒桶类型： 波本和雪利桶
泥煤含量等级： 20 ppm
限量版： 12,000瓶

100%艾雷第8版来自2008年和2012年装桶的23个波本桶和7个
欧罗索雪利桶。

50%ALC./vol.

闻香	清新，淡淡的柑橘味，带有一丝焦糖和雪利酒的甜味。
味觉	新鲜水果、香料和肉桂的完美平衡，以及婉转绵长的泥煤烟熏味。
余味	长久而干净的尾韵，平衡得很好。

CASK TYPE: BOURBON & SHERRY
PEATING LEVEL: 20 ppm
LIMITED EDITION: 12,000 BOTTLES

The 100% Islay 8th Edition is a vatting of 23 bourbon barrels and 7 Oloroso sherry butts filled between 2008 and 2012.

50%ALC./vol.	
NOSE	Fresh, light citrus notes with a hint of caramel and sherry sweetness.
TASTE	A lovely balance of fresh fruit, spice and cinnamon with soft lingering peat smoke.
FINISH	Well balanced with a long clean finish.

齐侯门苏特恩桶2018版 /
Kilchoman Sauternes Cask Finish 2018 Edition

酒桶类型：苏特恩桶
泥煤含量等级：50 ppm
限量版：10,000瓶

　　这款苏特恩桶威士忌的成品来自30桶2012年装桶的波本桶威士忌，在装瓶前，它们被转入苏特恩桶内熟成达5个月。

50%ALC./vol.

闻香	海洋泥煤烟熏味，柑橘和热带果香，带有黄油般的甜香。
味觉	柔软的奶油巧克力和甘草味，伴有阵阵泥煤烟熏味、肉桂和混合水果味。
余味	层层叠叠的泥煤烟熏味和果香，伴着持久的蜂蜜甜味。

CASK TYPE: SAUTERNES FINISH
PEATING LEVEL: 50 ppm
LIMITED EDITION: 10,000 BOTTLES

The Sauternes Cask Finish is a vatting of thirty 2012 bourbon barrels matured in Sauternes wine casks for five months before bottling.

50%ALC./vol.	
NOSE	Maritime peat smoke, citrus and tropical fruit with hints of buttery sweetness.
TASTE	Soft creamy chocolate and liquorice with waves of peat smoke, cinnamon and mixed fruit.
FINISH	Layered peat smoke and fruit with lasting honey sweetness.

拉加维林 / *Lagavulin*

酒厂照片 Distillery（照片由作者提供 Photo by author）

　　拉加维林蒸馏厂被认为是该国历史最悠久的酒厂之一，它坐落在艾雷岛南岸附近的一个小海湾，靠近丹尼维哥城堡遗址。1742年的时候，拉加维林这里至少有10个非法蒸馏炉，直到74年后由当地农民约翰·约翰斯顿组建为第一个合法蒸馏厂。

拉加维林蒸馏厂一些重要年份

　　1816年——当地农民及酿酒师约翰·约翰斯顿将这些建筑转变成一个合法的蒸馏厂并将其命名为拉加维林，这是该地区威士忌生产的首次合法运作。

　　1817年——第二个蒸馏厂出现了，被称为麦芽工厂，由一位叫阿奇博尔德·坎贝尔的人运作，后来被纳入了拉加维林。

　　1836年——约翰·约翰斯顿去世，亚历山大·格拉汉姆，一位格拉斯哥烈酒商，收购了蒸馏厂。

Lagavulin Distillery is thought to be one of the longest established distilleries in the country, and it is situated in a small bay near the south coast of Islay, close to the ruins of Dunyvaig Castle. In 1742, there were at least ten illicit stills at Lagavulin, and it wasn't until 74 years later that the first legal distillery was founded by local farmer John Johnston.

Some important years for Lagavulin Distillery

1816—Local farmer and distiller John Johnston converts the buildings into a legal distillery and names it Lagavulin, the first legal operation in the area.

1817—A second distillery appeared, known as Malt Mill, run by one Archibald Campbell. It is later subsumed by Lagavulin.

1836—John Johnston dies and Alexander Graham, Glasgow spirit merchant, acquires the distillery.

1861年——拉加维林蒸馏厂和农场的租约易手，落入詹姆斯·洛根·麦基公司手中，该公司由詹姆斯·麦基和格拉汉姆家族幸存的成员格拉汉姆上尉合作创立。

1878年——詹姆斯·洛根·麦基将他的侄子彼得·J.麦基带入威士忌生意。这是彼得第一次远行来到拉加维林，他之后还多次远道而来学习蒸馏威士忌的秘密。

1887年——詹姆斯·洛根·麦基公司接替了格拉汉姆公司。

1890年——彼得继任为高级合伙人，并且正是在他的领导下，拉加维林此后成为家喻户晓的名字。麦基在同事和下属之间更广为人知的是"待不住的彼得"，据说他的生活格言是"没有什么不可能"。这一年，公司改名为"麦基公司"。

1908年——麦基决定在原址修复两座建筑，这两座建筑被认为之前是在麦芽工厂名下的时候用作蒸馏屋以及仓库的。

1924年——麦基去世，麦基公司成为白马蒸馏者有限公司。一种名为S.S.风笛变奏曲的克莱德河"汽船"加入了运送大麦、煤炭和空桶到拉加维林的服务线，回程会带上装满拉加维林威士忌的酒桶。

1927年——白马蒸馏者和拉加维林加入了蒸馏者有限公司。

1939年——随着战争的爆发，女人们被征召入酒厂工作，直到1941年蒸馏厂最终在战争期内关闭。

1948年——电力被引入蒸馏厂。

1962年——"麦芽工厂"最终关闭，但这个传说中的威士忌的一份珍贵样品被安全地保存在拉加维林蒸馏厂里。

1974年——拉加维林关闭了它的麦芽发芽室，转而从埃伦港麦芽厂买进麦芽。

1861—The lease for Lagavulin Distillery and farm changes hands, falling under the control of James L. Mackie & Co., the company formed by James Mackie in partnership with the surviving member of the Graham family, Captain Graham.

1878—J. L. Mackie brings his nephew Peter J. Mackie into the business and Peter makes the first of many trips to Lagavulin to learn the secrets of distilling.

1887—James Logan Mackie & Co. succeed Graham & Co..

1890—Peter succeeds as senior partner, and it is under his guidance that Lagavulin will become a household name. Mackie is better known to colleagues and staff as "Restless Peter", said to live by the maxim "Nothing is Impossible". The name of the firm changes to Mackie & Co..

1908—Mackie decides to restore two buildings on site, believed to be former still house and store, to their former use, under the name of Malt Mill.

1924—Mackie dies and Mackie & Co. becomes White Horse Distillers Ltd.. *The S.S. Pibroch, a Clyde "Puffer"* enters service to transport barley, coal and empty casks to Lagavulin, returning with filled casks of Lagavulin.

1927—White Horse Distillers and Lagavulin join the Distillers Company Limited.

1939—With the advent of war, women are drafted in to work the distillery until 1941 when the distillery finally closed for the duration.

1948—Electricity is introduced to the distillery.

1962—Malt Mill finally closes, but a precious sample of this fabled whisky is kept safe at Lagavulin Distillery.

1974—Lagavulin closes its malting floors and buys in from Port Ellen Maltings.

拉加维林16年 / Lagavulin 16 Years Old

在橡木桶中陈酿至少16年，这一广受追捧的单一麦芽威士忌拥有艾雷岛南部典型的、厚重的泥煤烟熏味——同时也展现出一种让它变得着实有趣的干度。

43%ALC./vol.

闻香	风味强烈，泥煤烟熏味伴随着碘酒和海藻，以及一种丰富、深沉的甜味。
味觉	干燥泥煤烟熏味使口腔内充满一种柔和但强劲的甜味，跟着是海洋和盐味，略有木质味道。
余味	一个长而优雅的充满泥煤调的结尾，还带着大量的盐和海藻的韵味。

Lagavulin

Aged in oak casks for at least sixteen years, this much sought-after single malt has the massive peat-smoke that's typical of southern Islay—but also offering a dryness that turns it into a truly interesting dram.

43%ALC./vol.	
NOSE	Intensely flavoured, peat smoke with iodine and seaweed and a rich, deep sweetness.
TASTE	Dry peat smoke fills the palate with a gentle but strong sweetness, followed by sea and salt with touches of wood.
FINISH	A long, elegant peat-filled finish with lots of salt and seaweed.

拉加维林酒厂限量款 /
Lagavulin Distillers Edition

在佩德罗·西门内木桶中经历二重熟成，这是一款柔和芳醇的拉加维林酒、丰富的泥煤味和甜味，使人渴望更多。

43%ALC./vol.	
气味	强烈的泥煤和香草味。葡萄干的甜味抑制了烟熏味。碘酒味环绕着的泥煤味和爽口的、烤过的麦芽。可口且迷人。
口感	一种清澈的、带草木香的麦芽味，接着是泥煤田——烟熏味充满口腔。一种很咸的味道，中间是咖啡、香草和果香。
尾韵	即使用艾雷岛的标准来衡量也是难以置信的悠长。果香、泥煤味和持久的橡木味，耐人寻味。

Lagavulin

Double matured in Pedro Ximénez cask wood, this is a mellow Lagavulin, peat-rich, sweet and very more-ish.

43%ALC./vol.	
NOSE	Intense peat and vanilla. A raisin sweetness checks the smoke. Iodine-edged peat and crisp, roasty malt. Satisfying and enticing.
TASTE	A clear, grassy malt, then the peat lands—smoke filling the mouth. A very salty tang, the middle offers coffee, vanilla and fruit.
FINISH	Incredibly long, even by Islay standards. Fruit, peat and long-lasting oak. Very chewable.

拉弗格 / Laphroaig

酒厂照片 Distillery（照片由作者提供 Photo by author）

　　拉弗格的故事已有200多年，而它的故事可在一口啜饮中被道尽。拉弗格只与威士忌这种液体有关，不会——也不能——成为它自己以外的任何东西，它讲述了许多关于制造它以及饮用它的人、关于它所来自的这座岛对它的影响，以及关于围绕它的那片海的种种。拉弗格的故事是威士忌的故事，而威士忌是一种生活方式。

　　拉弗格最原始的东西很多遗失在那段在艾雷岛上生产威士忌是非法活动的年代里。结果就是，在1815年以前都没有关于拉弗格纳税的记录。然而，我们确实知道约翰斯顿家族至少从1776年开始就在奇戴顿地区进行农牧生产，并且，和大多数艾雷岛农民一样，会将多余的饲料——大麦酿造成威士忌。而且，至关重要的是，查尔斯和威利·多伊格被记录曾在拉弗格从事"一些酒厂的工作"。

The story of Laphroaig is over 200 years long, but it's one that can be told with one sip. Laphroaig is all about the liquid, a whisky that will not—cannot—be anything but itself; one that speaks volumes about the people who make and drink it, the influence of the island from which it comes, and the sea that surrounds it. The story of Laphroaig is the story of a whisky that is also a way of life.

Much of the origins of Laphroaig are lost to a time during which the making of whisky on Islay was illegal. As a result, there are no tax records pertaining to Laphroaig prior to 1815. However, we do know that that the Johnston family had been cattle farming the Kildalton area from at least 1776, and that, like most Islay farmers, would have turned excess feed—namely, barley—into whisky. Also, crucially, Charles and Willie Doig are recorded as having been commissioned to "do some work on a distillery" at Laphroaig.

仓库 Warehouse（照片由拉弗格提供 Photo by Laphroaig）

　　不管怎样，亚历山大和邓肯·约翰斯顿的威士忌的名声在这之前已经传开。基尔布赖德的水源非常好，而且当1825年法律更改的时候，之前一直作为副业的生意就开始了指数级增长。农场继续开到了20世纪60年代，而邓肯在1826年的税务账目中曾被列为酿酒师。拉弗格现在是一个完整建立起来的生意了，尽管是从当地领主奇戴顿的拉姆齐家租用的土地。

　　在接下来几十年过程中，这个酒厂几次被转手。唐纳德首先买下了亚历山大的全部股权，然后把它交给了另一个兄弟约翰和当地农民彼得·麦金太尔。麦金太尔一直照料着酒厂，直到他的儿子杜戈尔德在1857年足够年龄后接掌大权，取得更大成功。拉弗格在杜戈尔德的引领下继续扩张，尽管过程不乏千奇百怪的难题，既有内部的，也有外部的。当杜戈尔德1877年过世的时候，他的堂兄弟亚历山大·约翰斯顿(昵称"桑迪")接手过来，并在他妻子和妻妹的协助下管理这家酒厂。等到桑迪1907年过世的时候，酒厂传递给了约翰·约翰斯顿·亨特·约翰斯顿以及他的姐妹凯瑟琳和伊莎贝拉，也传给了他的姐夫威廉·S.亨特。在1908年，威廉和伊莎贝拉的儿子伊恩·亨特被任命为酒厂负责人。1923年，他获得酒厂的所有权。

Either way, the reputation of Alexander and Duncan Johnston's whisky preceded it. The water sourced from Kilbride was especially good, and when the law changed in 1825 so started the exponential growth of what had until then been something of a sideline business. The farm continued up until 1960s, and Duncan was mentioned as distiller in the excise accounts of 1826. Laphroaig was now a fully established business, albeit on land leased from the local laird, the Ramsays of Kildalton.

Over the course of the next few decades, the distillery changed hands several times. Donald first bought out Alexander, then left it in the care of another brother, John, and local farmer Peter McIntyre, who looked after it until his son, Dugald, was old enough to take up the reins in 1857. Success begat success. Laphroaig continued to grow under Dugald's guidance, though not without the odd hiccup, both internally and externally. When Dugald died in 1877, his cousin Alexander "Sandy" Johnston took over, and managed the distillery with the aid of his wife and sister-in-law. Sandy in turn died in 1907 and so the distillery passed into the hands of the wonderfully named John Johnston Hunter Johnston and his sisters Catherine and Isabella, and also to his brother-in-law William S Hunter. In 1908, William and Isabella's son Ian Hunter was appointed distillery manager. He would take up ownership of the distillery in 1923.

酒厂全景 Panoramic view of the distillery（照片由作者提供 Photo by author）

　　正是在伊恩的领导下，以及他的得力助手伊丽莎白·利奇·威廉姆森、昵称"贝西"的辅助下，拉弗格成了今天的模样。拉弗格不仅增加了蒸馏炉的数量，扩大了仓库和输出的规模，还在他们的任期内渡过了几次经济危机，从而买下了拉姆齐的房产；威士忌的生产得以稳定在了一个至今未变的配方上；并且，他带领这家威士忌酒厂走向全球，甚至能够卖给禁酒时期的美国。伊恩1954年去世，将酒厂留给了贝西。贝西一如既往地运营着酒厂——在需要的地方让其成长，扩张它的市场份额并且一直对亨特的秘密配方保密在心。她在1972年退休，并于10年后去世。当她还是酿酒师，并变得愈发重要的时候，贝西已在间歇地将她在拉弗格的股份出售给美国公司西格·伊万斯，在她退休那年卖掉了最后一部分。三年后，拉弗格被英国的韦博得集团收购，1985年，它又被接着转给了里昂联合集团，也叫联合多美集团。2005年，联合多美被保乐力加集团收购，并几乎同时，又被富俊公司收购，后者要将联合多美的非烈酒设施剔除，成立金宾公司。2014年，日本饮料公司三得利收购了金宾公司之后，拉弗格酒厂现在属于宾三得利。

It was under Ian's leadership, and that of his right hand woman, Elizabeth "Bessie" Leitch Williamson, that Laphroaig became the distillery it is today. Laphroaig not only expanded in terms of the number of stills, warehouse and output, but also over the course of their tenure survived several financial difficulties to purchase the estate of the Ramsays; settled on a recipe that hasn't changed to this day; and presided over a whisky that went global, managing even to sell in Prohibition America. Ian died in 1954, leaving the distillery to Bessie, who, true to form ran it exactly as before—growing it where necessary, expanding its market and always keeping the secret of the Hunter recipe close to her chest. She retired in 1972, and died 10 years later. While she was still distiller, and growth the priority, Bessie had intermittently sold her shares in Laphroaig to American company Seager Evans, the final block in the year she retired. Three years later, Laphroaig was bought by British Whitbread, after which it successively passed to Allied Lyons (1985), aka Allied Domecq, then on its acquisition of Allied Domecq to Pernod Ricard (2005), and then almost immediately to Fortune Brands, which was to strip away its non-spirits assets to form Beam Inc.. The distillery is now owned by Beam Suntory, following the Japanese drinks company's acquisition of Beam Inc. in 2014.

　　拉弗格在1994年得到查尔斯王子殿下授予的王室认证（英国王室供货许可证），同年，拉弗格之友俱乐部正式成立。查尔斯王子对酒厂当时的经理伊安·亨德森说："我希望你们继续保持传统的生产方法。我认为你们酿出了世界上最好的威士忌。"

酒厂鸟瞰图 Bird's-eye view of the distillery（照片由拉弗格提供 Photo by Laphroaig）

Laphroaig was granted a Royal Warrant by HRH Prince Charles in 1994, the same year in which it officially set up the Friends of Laphroaig Club. Prince Charles said to the then distillery manager Iain Henderson: "I hope you continue to use the traditional methods. I think you make the finest whisky in the world."

拉弗格珍选 / Laphroaig Select

拉弗格珍选得名于对美国和欧洲桶的特别挑选，选择的目的是它们在酒液熟成的过程中能赋予酒体的个性。全球酒款的灵感来自伊恩·亨特，最后一位掌管拉弗格的家族成员，也是第一个旅行到美国波本乡村的酿酒师，他发现了新的橡木桶来源，能够提供更一贯的品质以及新风味。他们已经运用了欧罗索雪利桶、美国白橡木桶（未灌注过波本威士忌）、佩德罗·西门内风干猪头桶、小容量木桶（又名：四分之一桶），最后当然还有首次灌注的波本桶。

40%ALC./vol.

闻香	先是泥煤味，然后是丰富的红色水果气息，略干。
味觉	最先感受到甜味，接着是经典的干、泥煤、灰烬的风味，跟着是一个丰富的尾韵。
余味	回味悠长，花香芬芳，最后是杏仁软糖和酸橙味。

Laphroaig

Laphroaig Select takes its name from the special selection of American and European casks, chosen for their unique character they give during maturation. The inspiration for Global expression comes from Ian Hunter, the last family member owner of the distillery, and one of the first distillers to travel to bourbon county in the USA to identify new sources of casks that would give him greater consistency as well as new flavours. They have used Oloroso sherry butts, straight American white oak (non-filled with bourbon), PX seasoned hogsheads, Quarter Casks and finally of course first-fill bourbon Casks.

40%ALC./vol.	
NOSE	Peat first, then ripe red fruits with a hint of dryness.
TASTE	Sweet up front then classic dry, peaty, ashy flavours followed by a rich finish.
FINISH	Long lingering and floral with marzipan and limes at the end.

拉弗格10年 / Laphroaig 10 Years Old

　　原始的拉弗格，在今天，还是用75年前伊恩·亨特发明的酿造方法来制作。在制作过程中，麦芽被燃烧的泥煤熏干。由这种仅在艾雷岛上才有的泥煤中产生的烟熏味，赋予了拉弗格独有的丰富风味。享用拉弗格这款10年威士忌的人们会首先注意到那大胆的烟熏味，跟着是一丝海草味以及令人惊讶的甜味。这个醇厚的酒款，是拉弗格所有酒款的基础，并且总是伴随着漫长的余韵。

40%ALC./vol.

闻香	厚重的烟熏味、海草味、"药味"，带着一丝甜蜜气息。
味觉	令人惊讶的甜味中带着一丝咸味，还有层层泥煤味。
余味	漫长而缠绵。

Laphroaig

The original Laphroaig, distilled the same way today as when Ian Hunter invented it over 75 years ago. In making Laphroaig, malted barley is dried over a peat fire. The smoke from this peat, found only on Islay, gives Laphroaig its particularly rich flavour. Those enjoying the 10 Years Old will first notice the bold, smoky taste, followed by a hint of seaweed and a surprising sweetness. This full-bodied variant is the foundation of all Laphroaig expressions and comes with a long finish.

40%ALC./vol.	
NOSE	Huge smoke, seaweedy, "medicinal", with a hint of sweetness.
TASTE	Surprising sweetness with hints of salt and layers of peatiness.
FINISH	Long and lingering.

拉弗格传说 /
Laphroaig Lore

由拉弗格酒厂经理约翰·坎贝尔制作，拉弗格传说从包括首次灌装的波本桶、小容量桶和欧罗索雪利桶中精选出来，创造出独一无二的丰富性。就如名字暗示的这般，这款新推出的拉弗格威士忌灵感正是源于酒厂经理、调酒师和匠人们世代相传的知识和传统。

48%ALC./vol.

闻香	丰富的烟熏气味带着海边的矿物、一丝灰烬和苦巧克力的味道。接下来是香草气息，伴随着油质的未曾烤过的栗子和带有麦芽甜味的乳脂软糖的气息。
味觉	一口辛辣一口丰富的泥煤味。
余味	短暂的干味之后是漫长的甜蜜回味。

Laphroaig

Crafted by Laphroaig Distillery Manager John Campbell, Laphroaig Lore is drawn from a variety of casks including first-fill bourbon barrels, quarter casks and Oloroso sherry hogsheads to create a unique richness. As the name suggests, this new offering from Laphroaig is inspired by the knowledge and traditions passed down from generation to generation of its distillery managers, blenders and craftsmen.

48%ALC./vol.	
NOSE	Rich and smoky with seaside minerals and a hint of ash and bitter chocolate drops. Vanilla follows with oily unroasted chestnuts and a hint of fudge with a malty sweetness.
TASTE	Richly peaty with a spicy chilli bite.
FINISH	Short dry finish and a long sweet aftertaste.

拉弗格四桶 / Laphroaig Four Oak

40%ALC./vol.

闻香	艾雷岛的麦芽泥煤烟熏味伴随着炖水果和温暖的烤香草的气息。
味觉	橡木烟熏味、海藻和光滑的奶油乳酪的味道。
余味	咸甘草根和泥煤味。

Laphroaig

40%ALC./vol.	
NOSE	Islay malt peat-smoke with stewed fruit and warm, toasted vanilla.
TASTE	Oak smoke and seaweed with smooth butter-cream.
FINISH	Salted liquorice and a peaty tang.

拉弗格三桶 / Laphroaig Triple Wood

48%ALC./vol.	
闻香	甜葡萄干和奶油杏仁、坚果风味，带土的泥煤篝火灰烬的气味。
味觉	更加浓郁的香草和水果风味，略带雪利酒的甜味。
余味	入口充盈，极长的回味被甜滑的焦糖味道平衡着。

Laphroaig

48%ALC./vol.	
NOSE	Sweet raisins and creamy apricots, nutty flavours, bonfire ash smell of the earthy peat.
TASTE	Creamier flavours of vanilla and fruit with just a suggestion of sherry sweetness.
FINISH	Mouth filling and extremely long, balanced by the sweet smooth caramel taste.

拉弗格大海 / Laphroaig An Cuan Mor

48%ALC./vol.	
闻香	无花果的丰富甜味，带有薰衣草花香以及黑胡椒味和天然蜂蜜的甜味。
味觉	非常辛辣的胡椒味，也有甘草根的味，略咸，发展为柔和的橡木单宁，可爱的橙皮味之后是干爽的泥土气息。
余味	奶油般的核桃味被拉弗格独特的泥煤味笼罩着。

Laphroaig

48%ALC./vol.	
NOSE	Rich sweetness of figs, with floral flavour of lavender, black pepper and sweet natural honey notes.
TASTE	Very peppery, spicy, also liquorice root with a slight saltiness. Moving to gentle oak tannins, lovely orange peel and then a dry earthiness.
FINISH	Creamy walnuts enveloped by the very distinct Laphroaig peatiness.

拉弗格10年桶强 /
Laphroaig 10 Years Old Cask Strength

55%-58%ALC./vol.

闻香	非常强劲，"药味"、烟熏气味、海藻和臭氧的特质堆叠成甜味。
味觉	大量泥煤味爆发开来，最后带着一丝甜味。
余味	漫长而美味。

Laphroaig

55%-58%ALC./vol.	
NOSE	Very powerful, "medicine", smoke, seaweed and ozone characters overlaying a sweetness.
TASTE	A massive peated burst of flavour with hints of sweetness at the end.
FINISH	Long and savoury.

拉弗格QA桶 / Laphroaig QA Cask

48%ALC./vol.

闻香	奶油的气味，伴随着太妃糖、坚果、山核桃、朗姆酒、葡萄干冰淇淋和橘皮味。
味觉	一记甜味的爆发，带着炽烈的香料味和少许糖浆的味道。
余味	中等长度，果香浓郁，有奶油冻味道和雪茄的烟雾味。

Laphroaig

48%ALC./vol.	
NOSE	Oily and buttery nose, with toffee, nuttiness, hickory, rum, raisin ice cream and zest.
TASTE	An explosion of sweetness, with fiery spice and a touch of treacle.
FINISH	Medium length, becoming fruity, with notes of custard and cigar smoke.

拉弗格PX桶 / Laphroaig PX Cask

48%ALC./vol.

闻香	无籽葡萄干和大葡萄干的雪利酒香气，略带甜甘草的香味。
味觉	泥煤味和橡木味以及更甜的重雪利风味。
余味	浓郁的泥煤味和浓厚的雪利桶橡木味，带着深沉的干味。

Laphroaig

48%ALC./vol.	
NOSE	Sherry aroma of sweet sultanas and raisins with a hint of sweet liquorice.
TASTE	Peat and oakiness and a sweeter heavy sherry flavour.
FINISH	Concentrated peat and thick sherried oak with a deep dryness.

拉弗格1/4桶 / Laphroaig Quarter Cask

48%ALC./vol.	
闻香	农家壁炉里泥煤燃烧的余烬，略带椰子和香蕉的香气。
味觉	深沉、复杂，有烟熏味，但用一种柔和的甜味惊艳了味蕾。
余味	非常长，并且干得恰到好处，带有烟熏和香料味。

Laphroaig

48%ALC./vol.	
NOSE	Burning embers of peat in a crofter's fireplace, hints of coconut and banana aromas.
TASTE	Deep, complex and smoky yet offers and surprises the palate with a gentle sweetness.
FINISH	Really long, and dries appropriately with smoke and spice.

拉弗格25年 / Laphroaig 25 Years Old

48.9%ALC./vol.

闻香	雪利酒的甜香伴随着历史悠久的艾雷岛泥煤味。一种顺滑的水果成熟香气，补充了背景中的盐的气味。
味觉	刚开始是一记泥煤味的爆发，被雪利酒的甜味安抚，之后发展为香辛苹果香。
余味	非常悠长且温暖，留下艾雷岛独特的味道。

Laphroaig

48.9%ALC./vol.	
NOSE	Sherry sweetness followed by the time-honoured Islay peat tang. A smooth fruit ripeness that complements the tang of salt in the background.
TASTE	An initial burst of peat restrained by a sherry sweetness that develops into spicy apple fruitiness.
FINISH	Very long and warming, leaving a distinct tang of Islay.

拉弗格30年 / Laphroaig 30 Years Old

53.5%ALC./vol.

闻香	柑橘和芒果气息辅以烤杏仁和椰子的气息。
味觉	油感，略带橙皮、香草、香菜叶和野人参的味道。
余味	长而复杂，带着徘徊不去、柔软多汁的核果类以及耐嚼的新鲜手卷烟叶的韵味。

Laphroaig

53.5%ALC./vol.	
NOSE	Mandarin and mango tempered by toasted almond and coconut.
TASTE	An oily mouth feel with hints of orange peel, vanilla, coriander leaf and wild root ginseng.
FINISH	Long and complex with lingering pulpy stone fruit and chewy green hand-rolled tobacco.

阿德纳霍 / **Ardnahoe**

酒厂商店 Distillery shop（照片由作者提供 Photo by author）

　　阿德纳霍酒厂是艾雷岛第九个也是最新的一个酒厂。它是莱恩家族长期以来的梦想，而家族已经拥有了亨特莱恩独立装瓶厂。该酒厂是100%的家族所有制。

　　阿德纳霍将生产一款经典的泥煤风格的艾雷岛单一麦芽威士忌，只用最好的原料制作，木质发酵槽、超长的林恩臂和虫管冷凝器。这在艾雷岛上是头一遭。

　　酿酒大师吉姆·马克伊万这样描述这款酒的几次测试："品质独特，质地和纯度着实令人惊叹"，并带有"经典的艾雷岛DNA"。

　　他们并不急于将他们的阿德纳霍酒液装瓶，并且不会打扰它，直到它完全准备好。

Ardnahoe Distillery is Islay's 9th and newest distillery. It is the long held dream of the Laing family who own Hunter Laing Independent Blenders & Bottlers. It's 100% family owned.

Ardnahoe will produce a classic peated style of Islay single malt, uniquely crafted from the finest ingredients, wooden washbacks, extremely long lyne arms and worm tub condensers. The very first on Islay.

Some test runs of the spirit have been described by Master Distiller Jim McEwan as "of a unique quality that is truly amazing in its texture and purity" with "classic Islay DNA".

They are in no rush to bottle their Ardnahoe spirit and will leave it until it is well and truly ready.

A.D.拉特雷——艾雷酒桶 /
A.D.Rattray—Cask Islay

自1868年以来，拉特雷一直是一个家族企业，从事苏格兰单一麦芽威士忌的调配和熟成。19世纪末，该公司的马车载着一箱箱上等葡萄酒和烈酒，经常出现在克莱德塞德和格拉斯哥市中心。

如今，安德鲁·杜瓦·拉特雷创建的公司位于苏格兰西海岸，由莫里森家族掌管，这个家族的名字和声誉都浸透在威士忌历史之中。

作为一家独立的威士忌公司，A.D.拉特雷现从苏格兰所有传统威士忌产区采购酒液和木桶。A.D.拉特雷桶系列和其他屡获殊荣的品牌Stronachie、艾雷酒桶、Bank Note已经出口到世界各地。

艾雷酒桶

A.D.拉特雷的艾雷酒桶是一款经典的单一麦芽酒，是赫布里底群岛女王的缩影。它的活泼足以满足泥煤爱好者，同时也足以吸引其他人来品尝他们第一口美味的艾雷岛麦芽酒。

艾雷酒桶在波本酒桶中熟成，泥煤度为轻量的35ppm。

A. D. Rattray has been a family business involved in the blending and maturation of Single Malt Scotch Whisky since 1868. The company's horse-drawn carts, carrying cases of fine wine and spirits, were a familiar sight on Clydeside and in Central Glasgow during the late 1800s.

Today, the company Andrew Dewar Rattray founded is based on the West coast of Scotland and in the hands of the Morrison family whose name and reputation are steeped in whisky history.

As an independent whisky company, A. D. Rattray now sources spirit and casks from all the traditional whisky regions of Scotland. The A. D. Rattray Cask Collection and award-winning brands, Stronachie, Cask Islay and Bank Note are exported throughout the world.

Cask Islay

A. D. Rattray Cask Islay is a classic single malt, epitomising the Queen of the Hebrides. It's lively enough to satisfy peat-lovers, yet inviting enough to tempt others to enjoy their first delicious dram of an Islay malt.

Cask Islay is matured in bourbon casks and admeasures a light 35 ppm.

艾雷酒桶 / Cask Islay

46%ALC./vol.

闻香	温暖炽热，像烤焦的泥炭窑，烟熏味后跟着散发出柑橘味。
味觉	油滑感、焦糖、多汁的大麦和丰富的泥煤烟熏味。
余味	烤熟的橙子，盐渍焦糖并以木头烟熏的味道作为背景。

46%ALC./vol.	
NOSE	Warm and fiery, roasting peat kiln, some citrus notes after the smoke.
TASTE	Oily, burnt toffee, juicy barley and rich peat smoke.
FINISH	Grilled orange, salted caramel and wood smoke in the background.

奥歌诗丹迪——冒烟的乔治 /
Angus Dundee—Smokey Joe

英国奥歌诗丹迪酒厂总部位于伦敦，1950年由邦史都华前高管特里·希尔曼创建，是一家苏格兰威士忌调和公司，现在由他的两个孩子塔尼亚和亚伦经营。

2000年，奥歌诗丹迪收购了JBB（大欧洲）已沉寂的托明多酒厂，获得了单一麦芽威士忌的供应。托明多是战后新一代功能齐全的苏格兰酒厂之一，始建于1964年。尽管外观简朴，但它位于一个美丽而遥远的乡村，距离格兰威特酒厂只有几英里。

在收购托明多三年以后，奥歌诗丹迪从联合多美集团手中收购了格兰卡登，以增加酒液供应。与托明多相比，格兰卡登有着悠久的威士忌酿造历史，由乔治·库珀于1825年创立。

除了两家酿酒厂，奥歌诗丹迪还在格拉斯哥附近的科特布里奇拥有一家装瓶厂。

冒烟的乔治

冒烟的乔治装瓶酒精度为46%，以增强冲击力和风味。组成冒烟的乔治的威士忌酒液皆是香气高度迷人、烟熏味和辛辣口感平衡以及回味悠长之品。

Based in London, Angus Dundee Distillers Plc. was established in 1950 by ex-Burn Stewart executive Terry Hillman as a Scotch whisky blending company, and is now run by his children Tania and Aaron.

In 2000, Angus Dundee purchased the silent Tomintoul Distillery from JBB (Greater Europe), giving it access to single malt stocks. Tomintoul was one of the new breed of functional post-war Scottish distilleries, established in 1964. Despite its austere appearance it was located in a beautiful if remote country, just a few miles from the great Glenlivet Distillery.

Three years after buying Tomintoul, Angus Dundee acquired Glencadam from Allied Domecq in order to augment spirit supply. By contrast with Tomintoul, Glencadam boasted a long history of whisky-making, founded in 1825 by George Cooper.

In addition to its two distilleries, Angus Dundee also owns a bottling plant at Coatbridge, near Glasgow.

Smokey Joe

Smokey Joe is bottled at 46% ABV to enhance impact and flavour. The whisky that makes up Smokey Joe has been selected for its highly engaging nose, balance of smoke and peppery spice on the palate and a lingering finish.

冒烟的乔治 / Smokey Joe

46%ALC./vol.

闻香	迷人的烟熏混合着精致的海洋气息和柑橘味。
味觉	烟熏味和辛辣香料味的微妙平衡。
余味	绵长的香料以及泥煤气息。

Smokey Joe

46%ALC./vol.	
NOSE	Engaging smoke mixed with refined marine hints and citrus notes.
TASTE	Exquisite balance of smoke and peppery spice.
FINISH	Lingering spice and peat.

道格拉斯莱恩——重泥煤 / Douglas Laing—Big Peat

　　道格拉斯莱恩创立于1948年，是世界上高端苏格兰威士忌的领先制作者和供应商，主要生产手工威士忌、小批量威士忌和单桶威士忌。现为家族企业的第三代。

　　在2017年，包括重泥煤在内的一系列道格拉斯莱恩威士忌获得了世界上最大、最值得信赖的食品和饮料大奖。

Established in 1948, Douglas Laing is a leading creator and purveyor of the finest Scotch Whisky with particular focus on artisan, Small Batch and Single Cask bottling. The family business is in its third generation.

In 2017, a range of Douglas Laing whiskies, including Big Peat, received the Great Taste Award, the world's largest and most trusted food and drinks award.

卡拉·莱恩、克里斯·莱格特、弗雷德·莱恩以及家养犬库珀
Cara Laing, Chris Leggat, Fred Laing and Cooper family dog
（照片由道格拉斯·莱恩提供 Photo by Douglas Laing）

重泥煤 / Big Peat

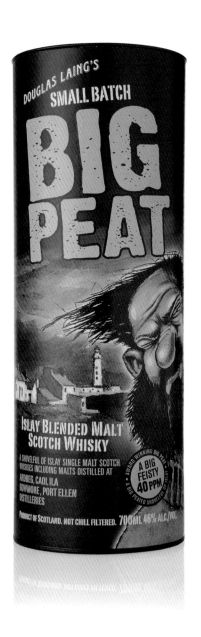

泥煤含量等级： 40 ppm

46%ALC./vol.

闻香	新开瓶，气味偏咸且干净，进而发展为麦芽在泥煤上烘干的味道。一种潮湿的、朴实的特质，迅速给人强力的冲击。
味觉	你会联想到灰烬、甜焦油、海滩以及冒烟的烟囱。
余味	绵长的尾韵，咸味、强烈的甘草味、烟熏味和更多的篝火灰烬和酚质气息。

PEATING LEVEL: 40 ppm

46%ALC./vol.	
NOSE	Opens fresh, salty and clean, and develops to malted barley dried over peat. A damp, earthy character that's immediately punchy.
TASTE	On the palate you'll detect ashes, sweet tar, beaches and smoking chimneys.
FINISH	A long and lingering finish with salty, tangy liquorice, smoke and yet more bonfire ashes and phenolic quality.

重泥煤2018年圣诞节版 / Big Peat Christmas 2018
（限量版 LIMITED EDITION）

它是许多单一麦芽威士忌的混合体，包含来自阿贝、卡尔里拉、波摩和埃伦港生产的单一麦芽威士忌。

53.9%ALC./vol.

闻香	浓重烟熏味和沿海气息，夹杂着菠萝味。
味觉	灰烬朴实的橡木味，带着层层的烹制香料和煮开的甜点味。
余味	绵延消失的煤烟和白胡椒的味道。

A shovelful of Islay Single Malt Scotch Whiskies including Malts distilled at Ardbeg, Caol Ila, Bowmore and Port Ellen.

53.9%ALC./vol.	
NOSE	Dense smoke and coastal air, punctuated by hints of pineapple.
TASTE	Ash and earthy oak, with layers of cooking spice and boiled sweeties.
FINISH	Lingering hints of soot and white pepper.

重泥煤2018年艾雷岛嘉年华版 /
Big Peat Fèis Ìle 2018 Edition

限量版：500箱

48%ALC./vol.

闻香	最初是干海藻，然后是烤肉以及糖浆的气息。
味觉	烟熏大麦以及潮湿的皮革，烟囱里的烟灰以及一丝丝柠檬皮的味道。
余味	绵长且令人回味，带有海盐味，加上培根以及烧焦橡木的味道。

LIMITED EDITION: 500 CASES

48%ALC./vol.	
NOSE	Initially dried seaweed then comes BBQ meats and golden syrup.
TASTE	Smoked barley with damp leather, chimney soot and a hint of lemon rind.
FINISH	Long and moreish with a sea-salt tang plus streaky bacon and charred oak.

圣水蒸馏者 / Elixir Distillers

　　圣水蒸馏者是烈酒的创造者、调和者和装瓶商。他们的专长首先是苏格兰威士忌，同时还有爱尔兰、日本和美国的威士忌，以及朗姆酒和龙舌兰酒。

　　其核心品牌是阿斯凯格港、艾雷元素、苏格兰单一麦芽威士忌和黑托特。此外，他们还有其他几个品牌正处在开发中。目前，他们向20多个国际市场出口品牌，并对这些品牌呈陡坡式增长感到欣喜。团队的每一位成员都是烈酒爱好者，拥有广阔的产品知识，期待着创作出世界上的下一款佳酿。

阿斯凯格港

　　自2009年开始装瓶，所以今年是第十一个年头。

　　阿斯凯格港是一个艾雷岛的单一麦芽威士忌系列，体现了艾雷岛及其人民的独特精神。它的特性中汇聚了来自这个迷人的岛上的浓烈烟熏味和柔和的果香。

　　该系列以吸引最挑剔的威士忌鉴赏家为开发方向，同时又吸引了新手威士忌饮家。

　　他们选择了独特的艾雷岛单一麦芽威士忌，并且创建了一个他们相信会被公认为艾雷岛经典的系列。

　　系列内的每一款酒都是限量装瓶，尽管意识到每款装瓶会有不同，但这么做的目的在于，随着时间的推移实现威士忌质量和特性的一致性。为保证每一瓶威士忌都维持它原有的风味和个性，这些威士忌未经冷凝过滤，也未添加任何色素。

Elixir Distillers is a creator, blender and bottler of fine spirits. Their expertise is primarily Scotch whisky, together with Irish, Japanese and American whiskies as well as rum and Tequila.

The core brands are Port Askaig, Elements of Islay, Single Malts of Scotland and Black Tot. In addition, they have several other brands in development. Currently they export brands to more than 20 international markets and are delighted that all are growing on a steep upward curve. Every member of team is a spirits enthusiast with immense product knowledge, looking to produce the world's next great drink.

Port Askaig

Bottling began in 2009 so this year is the 11th anniversary.

Port Askaig is a range of Islay single malt whiskies that embodies the unique spirit of Islay and its people. Its character brings together the robust smokiness and soft fruitiness found in this beguiling island.

The range has been developed to appeal to the most discerning of whisky connoisseurs while also appealing to the novice whisky drinker. They have selected exceptional casks of Islay single malt and created a range that believe they will become recognised as an Islay classic.

Each expression within the range is bottled in limited batches. While they recognises that each bottling will vary, the aim is to achieve a consistency of quality and character over time. To ensure each whisky maintains its original flavour and character, the whiskies are not chill-filtered and no colouring is added.

艾雷元素

艾雷元素是一个标志性的、小批量的、专业来源的独立瓶装威士忌系列。限量版系列包括了来自各个艾雷岛独立酿酒厂装瓶的单一麦芽威士忌，它们由周期符号和批号（如Lp1、Ar2等）作为识别。它们可以从单桶来源装瓶，也可以每次挑3到20桶精选的酒作为来源，小批量装瓶。这些酒款既赞扬了来源酒厂的独特品质，又将它们呈现为十足平衡和有特色的威士忌。

Elements of Islay

The Elements of Islay is a range of iconic, small-batch, expertly sourced independently bottled whiskies. The Limited Edition range consists of single malt bottlings from individual Islay distilleries identified with periodic symbols and batch numbers (Lp1, Ar2, and so on). They are bottled either as single casks or as small batches between 3 and 20 carefully selected casks. These expressions celebrate the distilleries' individual character while presenting a balanced and characterful whisky bottled at full proof.

阿斯凯格港100° Proof / Port Askaig 100° Proof

57.1%ALC./vol.

闻香	丰富，带着肉香的烟熏味与它的浅色调形成反差，新鲜煮熟的火腿以及燃烧着的原木，加入了柔和的海岸特征：海雾和多石的岩坑水洼。
味觉	开始是锋利的带着石质的味道，强烈的矿物感，很快就变成甜而浓郁的牛奶咖啡、牛奶巧克力和香辛肉桂硬糖味。香料味逐渐增强，加入甘草和肉桂炖苹果，以及浓重的木炭烟味。
余味	甜、辛辣且持久。肉桂巧克力韵调慢慢褪色成干苹果皮的味道。

Port Askaig

57.1%ALC./vol.	
NOSE	Rich, with a meaty smokiness that belies its light colour. Freshly cooked ham and burning logs are joined by a soft coastal character of sea spray and stony rock pools.
TASTE	Sharp and stony initially, with an intense minerality that quickly becomes sweet and rich—milky coffee, milk chocolate and spicy cinnamon gobstoppers. The spice grows, adding liquorice and stewed apple with cinnamon, along with heavy charcoal smoke.
FINISH	Sweet, spicy and very long. Cinnamon-spiced chocolate slowly fades to drying apple skin.

阿斯凯格港陈酿8年 / Port Askaig Aged 8 Years

45.8%ALC./vol.

闻香	清新、干净，充满活力的柠檬油和海雾的气息。烟熏味温和而诱人，使人联想到在海边篝火中逐渐冷却的木柴余烬和煤炭。
味觉	浓郁的柑橘味和盐烤鲭鱼的味道，经典的艾雷岛泥煤味混合了金银花的微甜，背景中的草本植物的特色增加了平衡感。
余味	随着烟灰和烟熏味散尽并准备再喝一口时，麦芽糖的味道仍绵延不散。

Port Askaig

45.8%ALC./vol.	
NOSE	Fresh, clean and vibrant with notes of lemon oil and sea spray. The smoke is gentle but alluring, reminiscent of wood embers and coal, cooling on a beach fire.
TASTE	Rich citrus fruit and salt baked mackerel. Classic Islay peat reek is mixed with a subtle honey suckle sweetness while a background herbal character adds balance.
FINISH	Barley sugar lingers as the soot and smoke dissipate leaving the palate clean and ready for another sip.

阿斯凯格港陈酿14年（波本桶）/
Port Askaig Aged 14 Years (Bourbon Cask)

45.8%ALC./vol.

闻香	炖水果、烧烤煤炭、海雾以及海岸岩石味。甜烟消散，新鲜柑橘和咸味焦糖的核心味道崭露头角，结合了经典的艾雷岛风味和清新的水果味。
味觉	烧焦的蜂窝和烤过的葡萄干，灌入了五香料和泥煤烟熏味。水果蜜饯和冰冻柠檬果子露作为底层基调，给人一种极其丰富多汁的口感，还带着焦糖布丁般的香味和美味的泥煤调子的焦糖层次。
余味	持久浓重，有枫糖腌制的培根和利口酒浸泡的大葡萄干的风味。一个非常漂亮清新且带着泥煤味的结尾。

Port Askaig

45.8%ALC./vol.	
NOSE	Stewed fruits and barbecue coal sea spray and coastal rocks. Sweet smoke dissipates as core flavours of fresh citrus and salted caramel come to the fore combining classic Islay flavours with refreshing fruity notes.
TASTE	Burnt honeycomb and baked raisins infused with five-spices and peat smoke. Candied fruits and sherbet lemon build as an underlying earth peat, bringing together a superbly rich and juicy mouthfeel, with crème-brulee-like notes and delicious peat-infused caramel layer.
FINISH	Long and heavy with maple-cured bacon and liqueur-soaked sultans. A beautifully fresh and peaty finish.

阿斯凯格港陈酿19年 / Port Askaig Aged 19 Years

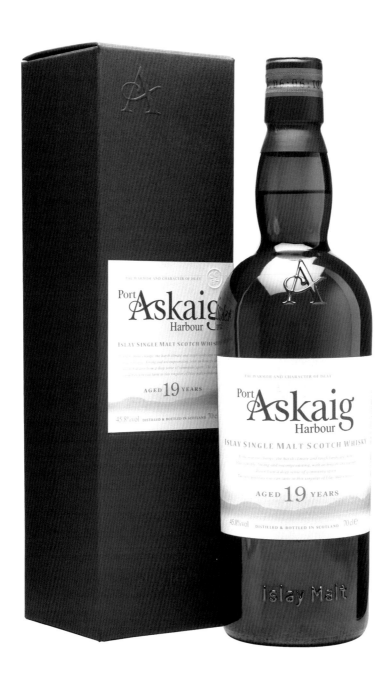

45.8%ALC./vol.

闻香	清新的海洋气息，加上海边的篝火和泥泞的泥煤气息。酸柠檬和苦橙编成协奏曲，带着甜甜的海绵蛋糕面糊味、花香和五味粉的气息使之达到平衡。
味觉	糖浆和甜味作为开始，美味的烟熏味随即滚滚而来——灰烬和泥土般，带着些微焦桶的甜味和木炭的干燥。散发着花香、辛辣和甜蜜的气息，小豆蔻、丁香、果子露喷泉（一种糖果）、蜜饯柠檬和紫罗兰的芬芳交织在一起。
余味	悠长绵延，泥土的芬芳，接着被热带水果和柑橘汁的味道所取代。

Port Askaig

45.8%ALC./vol.	
NOSE	Crisp and maritime, with a combination of seaside bonfires and muddy peat. Sour lemons and bitter oranges provide a counterpoint, with sweet sponge-cake batter, floral notes and ground spice giving balance.
TASTE	Syrupy and sweet to start, with savoury smoke rolling in soon after—ashy and earthy, with touches of barrel-char sweetness and charcoal dryness. Floral, spicy and sweet notes develop, with cardamom and clove joined by Sherbet Fountains, candied lemons and a hint of violets.
FINISH	Long and lingering, with earthiness giving way to tropical fruit and citrusy sap.

阿斯凯格港陈酿30年 / Port Askaig Aged 30 Years

51.1%ALC./vol.

闻香　挤满了各种风味——这里有很多反应正在进行。浓郁的蜂蜜甜味，背景是花香、白甜瓜、麦芽糖和香蕉味——新鲜而成熟，还有棉花糖的气味。与之相随的是期待中的艾雷岛泥煤烟熏味，尽管它在这儿是克制和优雅的。

味觉　从鼻腔中提取香气，在此基础上建立了丰满而宏大的口感。烟熏味更加明显，丰富的湿泥煤和炭火增加了甜和干燥的层次并补充了水果风味，并允许柑橘香交织呈现——蜂蜜和柠檬的基调中夹带一丝药水味。

余味　刚开始激烈，慢慢褪色成一种绵长的烟熏感，衬托在口腔内的水果味之上。

Port Askaig

51.1%ALC./vol.	
NOSE	Packed with flavour—there's lots going on here. Rich honey sweetness, with background floral notes, white melon, barley sugar and bananas—fresh and ripe, and foam sweets. With the expected Islay peat smoke, though it is restrained and elegant.
TASTE	Full and generous, taking the flavours from the nose and building on them. The smoke is more evident, with rich wet peat and coal fires adding a sweet and dry layer that complements the fruit and allows a citrus thread to emerge—honey and lemon with a hint of the medicinal.
FINISH	Initially intense, but slowly fading into a long, lingering sootiness, backed up by the palate's fruitiness.

阿斯凯格港陈酿45年 / Port Askaig Aged 45 Years

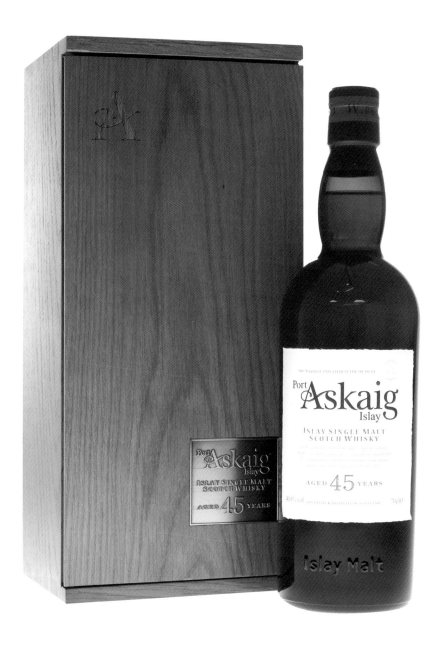

40.8%ALC./vol.

闻香	浓厚，水果香；热带水果和庄园果香缠绕着，伴随苹果、梨与未成熟的芒果、杨桃和木瓜的气味。柔和的酸度被犹如蜂蜜和花蜜的甜味以及黏性所平衡。
味觉	丰富，脐橙苦甜参半，佐以多肉芒果和温暖香料的气息。与之形成鲜明对比的是草本特征的味道，绿茶和薄荷叶提供了清新的自然气息。
余味	悠长，水果味转化成为草本的味道，定格在一个新鲜、凉爽的薄荷糖和薄荷醇的尾韵上。

Port Askaig

40.8%ALC./vol.	
NOSE	Rich, thick and fruity; tropical and orchard fruit intertwine, with apples and pears sat alongside unripe mango, star fruit and papaya. The gentle acidity of the fruit is balanced by sweet and sticky notes of honey and nectar.
TASTE	Rich, navel-orange bittersweetness is joined by fleshy mangoes and warming spice. With that is a contrasting herbal character, with green tea and mint leaves providing fresh and green notes.
FINISH	Long, with fruit becoming herbal before settling on a fresh and cooling mint and menthol end.

艾雷元素 Ar$_{10}$ /
Elements of Islay Ar$_{10}$

　　由两桶2001年蒸馏陈于波本桶中的酒为来源装瓶，这个阿贝旗下的酒款是甜甜的、烟熏味的，浓郁而精致。酒厂的经典矛盾性永远存在于最新装瓶的Ar系列中，使它非常受人尊敬。

52.4%ALC./vol.

闻香	甜香草和白巧克力与木质烟熏和碘酒的完美融合。
味觉	更多的白巧克力饼干、甜焦糖和果园水果（烤梨和苹果）伴随着强烈的坚果味，经典的篝火烟雾和木炭提醒我们这款威士忌的起源。
余味	山核桃派和烟熏杏仁味。

Elements of Islay

A vatting of two bourbon barrels distilled in 2001, this Ardbeg is sweet and smoky, rich yet delicate. The classic contradictions of the distillery that makes it so revered are ever-present in the latest Ar bottling.

52.4%ALC./vol.	
NOSE	Sweet vanilla and white chocolate mix perfectly with wisps of woody smoke and iodine.
TASTE	More white-chocolate cookies, sweet caramel and orchard fruit (baked pears and apples) sit alongside an intense nuttiness with classic bonfire smoke and charcoal reminding us of this whisky's origin.
FINISH	Pecan pie and smoked almonds.

艾雷元素 Bn$_7$ / Elements of Islay Bn$_7$

蒸馏于2001年，之后超过16年在一对西班牙欧罗索雪利酒桶中熟成。充满了浓郁的焦糖和黑巧克力风味——一个完美的冬季暖物。世界上仅生产1620瓶。

55.7%ALC./vol.

闻香	一锅融化的牛奶巧克力粒，中间还有焦糖。烤肉桂棒、姜粉、大葡萄干和核桃——完美的冬季蛋糕组合。
味觉	厚而柔滑的质地，有牛奶巧克力和糖霜樱桃的味道。经典的苦橙果酱涂在刚烤好的全麦面包上。混合香料紧跟在后：葛缕子籽、丁香和多香果。
余味	多汁、红色浆果、黑加仑甜酒、肉桂香料以及挥之不去的柑橘气息，类似蜜饯橘子皮。

Elements of Islay

Distilled in 2001 and aged for more than 16 years in a pair of Oloroso sherry butts. Full of rich caramel and dark-chocolate flavour—a perfect winter warmer. Only 1620 bottles were produced.

55.7%ALC./vol.	
NOSE	A pan of melting milk chocolate buttons with caramel centres. Toasted cinnamon sticks and ground ginger mixed with sultanas and walnuts—the perfect winter cake mix.
TASTE	A thick and creamy texture with tones of milk chocolate and glacé cherries. Classic bitter-orange marmalade spread on freshly-toasted granary bread. Mixed spice follows: caraway seeds, cloves and allspice.
FINISH	Juicy with red berries, blackcurrant cordial, cinnamon spice and a lingering citrus note akin to candied orange peel.

艾雷元素 Cl$_{11}$ / Elements of Islay Cl$_{11}$

这款威士忌蒸馏于2008年，最后花了18个月的时间在两个来自索莱拉的雪利桶中熟成。古老的木材散发出一种微尘的辛辣气息，与卡尔里拉原酒中的盐和海水味相互映衬。全世界仅生产1206瓶。

55.4%ALC./vol.

闻香	黑巧克力，木炭烟雾和黄油黑加仑。荨麻和接骨木花更为清新自然的香气平衡了冬季特饮的辛香。
味觉	口感厚重有密度感，在厚厚的甜酱中裹着丰富的莫利洛樱桃，酸甜的完美平衡。泥煤味是草本基调，让人想起陈年的荨麻酒，随后是苦涩的巧克力味。
余味	可可豆和阿布罗斯熏鱼的味道挥之不去。

Elements of Islay

This whisky was distilled in 2008 and spent the last 18 months of maturation in two sherry butts taken from Solera. The old wood has imparted a dust spicy note that plays well against the salt and brine found in Caol Ila's spirit. Only 1206 bottles were produced.

55.4%ALC./vol.	
NOSE	Dusty dark chocolate, charcoal smoke and buttered currants. Winter-punch spice is balanced by fresher, greener notes of nettle and elderflower.
TASTE	Mouth coating and dense, with rich Morello cherries in a thick sweet sauce—the perfect balance of sweet and sour. The peat is herbaceous, reminiscent of aged Chartreuse, and is followed by notes of bitter chocolate.
FINISH	Cocoa nibs and Arbroath smokies linger.

艾雷元素 Lg$_8$ / Elements of Islay Lg$_8$

蒸馏于2006年，陈酿于两个波本桶中，第八个装瓶的拉加维林佳酿融合了煤烟、烤肉和甜水果的所有经典风味。是一种令人愉悦的提神佳饮。

59.5%ALC./vol.

闻香	焦糖的甜蜜香气，芒果和香辛料呼应烟熏味基调，如同冷却的火炉。
味觉	一股烟熏味马上覆盖了微妙的香气，但消失得很慢，取而代之的是甜的焦糖芒果和桃子，与咸的烧焦的烤火腿的味道混合在一起。背景中隐约可品出八角和丁香的味道。
余味	清新的咸味萦绕唇齿间，伴着甜美的烟味和淡淡的碘酒味。

Elements of Islay

This whisky was distilled in 1998 and bottled from a single Pedro Ximénez sherry butt. The balance of rich fruit and smoke is perfect and the signature notes of 1990s' Laphroaig distillate are present and correct. Only 714 bottles were produced.

54.3%ALC./vol.	
NOSE	Coal fires, candy floss and lobster pots. Thick, dense smoke combines with notes of kirsch. The peat becomes more medicinal, showing classic distillery character. Darker, almost burnt-tyre notes remain.
TASTE	Juicy, with rich stewed fruit and smoked salt. Well balanced and reminiscent of perfectly cooked game meat covered in a port and redcurrant sauce. Notes of beach BBQ, charred mackerel and sweet langoustines develop, reminding us of a perfect summer day on Islay.
FINISH	Smoked sea salt and dark fruit linger, leaving the palate slightly dry and ready for a second sip.

艾雷元素 Oc5 /
Elements of Islay Oc5

　　泥煤怪兽回来了，带回它经典的麦片气息，一些多汁水果味为它锦上添花。这款来自奥肯特摩的威士忌蒸馏于2011年，并陈酿于一个波本桶中。

59.8%ALC./vol.

闻香	层叠、迷人，香味中有大量麦芽、潮湿的稻草和泥炭烟味，还有葡萄和橙子的多汁水果沙拉的香味。
味觉	先是更多的水果与坚果、太妃糖以及谷物混合在一起，然后一股烟熏感绽开变为主调，慢慢消散，最后形成烤苹果派的味道。
余味	烤苹果和从烧柴的炉子上冒出的一缕烟雾的韵味。

Elements of Islay

The peaty beast returns, bringing its classic cereal notes and some juicy fruit for good measure. This Octomore was distilled in 2011 and aged in a single bourbon barrel.

59.8%ALC./vol.	
NOSE	Multi-layered and fascinating, the nose offers lots of malty cereal, wet straw and peaty smoke alongside juicy fruit-salad notes of grapes and oranges.
TASTE	More fruit at first mingles with nuts, toffee, and cereal before an explosion of smoke takes over, slowly dissipating and leading to baked apple pastries.
FINISH	Baked apples and wisps of smoke from a wood-burning stove.

艾雷元素 Pl₅ /
Elements of Islay Pl₅

蒸馏于2009年，这一批酒液是再次灌注的猪头桶和初次灌注的桶中熟成的酒液的结合。这是典型的夏洛特港的酒体，烟熏味浓厚，泥煤味偏甜——就像走过农家庭院的篝火旁闻到的味道。全世界只生产了582瓶。

63.1%ALC./vol.

闻香	甜烟草和农场的特质，这是典型的顶级夏洛特港的酒。松木和烟熏橡木片，随之而来的是海浪和柏油绳的气息。
味觉	开始时精致而简朴。柠檬油和木本香料的基调在杯中形成，变得更加丰富。浓烈的烟味在烤熟的佐了海鲜沙司的白菜周围翻滚。香草的甜味和暗淡的烟味互相竞争，产生了更深更暗的复杂性。
余味	干爽而微妙，有瓶龄意大利苦艾酒、艾草、丁香和肉桂皮的味道。

Elements of Islay

Distilled in 2009, this batch is a vatting of a refill hogshead and a first-fill barrel. It is classic Port Charlotte, with dense smoke and sweet peat—like walking past a farmyard bonfire. Only 582 bottles were produced.

63.1%ALC./vol.	
NOSE	Sweet tobacco and the farmyard character that's typical of the best Port Charlottes. Pine wood and smoky oak chips follow accompanied by sea spray and tarred rope.
TASTE	Refined and austere to start with. Lemon oil and woody spice develop in the glass, becoming richer. Dense and aromatic smoke billows around charred pak choi in hoisin sauce. Vanilla sweetness and dirty smoke play against each other, creating deeper and darker complexity.
FINISH	Dry and subtle with hints of bottle-aged Italian vermouth, wormwood, cloves and cassia bark.

伊恩·麦克劳德蒸馏者——烟头 /
Ian Macleod Distillers—Smokehead

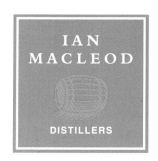

老莱昂纳多·J.罗素，公司的创始人，坚信独立的价值。不受制于任何一个酒商，他只采购那些符合他自己和顾客严格标准的威士忌。他和他的顾客显然都很有品位，因为80多年一路走来，伊恩·麦克劳德积累了令人羡慕的优质烈酒组合，旗下拥有格兰哥尼、特姆杜、斯凯岛、烟头苏格兰威士忌和爱丁堡杜松子酒等品牌。作为世界第十大苏格兰威士忌公司，他们目前每年生产和销售超过1500万瓶烈性酒。

伊恩·麦克劳德是买主自有品牌市场的主要供应商，以创新思维著称。伊恩·麦克劳德还为欧洲一些大型超市集团提供自有品牌的烈酒，超过40年。

随着持续在装瓶生产线、在布里斯本装瓶厂、在仓储以及最重要的无可匹敌的苏格兰威士忌的库存上投资，伊恩·麦克劳德蒸馏者比以往任何时候都更有能力满足全球对其产品日益增长的需求。

关于烟头

◎ 烟头是一款艾雷岛单一麦芽威士忌，被伊恩·麦克劳德赋予了一个大胆的新造型。

◎ 于2006年首次推出，烟头被贴上了单一麦芽威士忌中的狂野者之标签。

◎ 屡获殊荣的优质威士忌，以态度为主导的定位和重泥炭口味。

Leonard J Russell Snr, company's founder, was a firm believer in the value of independence. Beholden to no single distiller, he bought only those whiskies which met his own and his customers' exacting standards. Both clearly had good taste, because more than 80 years down the line, Ian Macleod Distillers has built up an enviable portfolio of premium quality spirits and is a proud brand proprietor of Glengoyne, Tamdhu, Isle of Skye, Smokehead Scotch whiskies and Edinburgh Gin to name but a few. The world's 10th largest Scotch whisky company, they currently produce and sell over 15 million bottles of spirits every year.

A major supplier to the Buyer's Own Brand market, with a reputation for innovative thinking, Ian Macleod Distillers has also supplied own-label spirits to some of Europe's largest supermarket groups for over 40 years.

With continued investment in bottling plant, Broxburn Bottlers, in warehousing and most importantly of all, their unrivalled stocks of Scotch whisky, Ian Macleod Distillers is in a stronger position than ever to meet the increasing demand for its products from around the world.

About Smokehead

◎ Smokehead is an Islay Single Malt Whisky and has been given a bold newlook by Ian Macleod Distillers.

◎ Originally launched in 2006, Smokehead has been labeled the wild one of Single Malt Whisky.

◎ Award-winning premium whisky that has an attitudeled positioning and heavily peated taste.

烟头 / Smokehead

43%ALC./vol.

第一印象	仿佛有什么强有力以及炽热的芬芳之物在等着你。厚重木烟，丰富朴实的泥煤。香辛和甜蜜的极致。新鲜柠檬，爽口的姜，浓郁的李子果酱。
完全的风味	在烟熏味再次冒出来之前，一记炸开的惊人胡椒味、泥煤的热度，被蜂蜜的甜味所抚平。
尾韵	异国情调的香料和奇特的强烈柑橘味都把你带入一种虚假的平静，同时泥煤味狂啸着折回来，再次击中你的感官。

IAN
MACLEOD

DISTILLERS

43%ALC./vol.	
The First Impression	The aroma of something powerful and fiery awaits you. Thick, heavy woodsmoke. Rich, earthy peat. Extremes of spice and sweetness. Fresh lemon, zesty ginger, rich plum jam.
The Full-on Flavour	An explosion of breathtaking peppery, peaty heat, soothed by honeyed sweetness, before the smoke comes to the fore again.
FINISH	Exotic spices and a curious citrus tang of mandarin, both lulling you into a false sense of calm as the peat roars back to hit your senses again.

烟头高压 / Smokehead High Voltage

58%ALC./vol.

第一印象	闻一闻，感受这款烈酒对感官的冲击及鼻子的刺痛。强烈的泥煤和烟熏、海上的空气、麦片粥、干净的香草和太妃糖的味道。
完全的风味	烈性酒的灼烧感，口腔充满烟熏味，感觉厚得几乎有油，之后软化并变得更偏向奶油感，带有坚果、海水和柑橘的气味。
尾韵	嘴巴几乎处于休克状态，麻木和刺痛。甜味和慢慢消退的烟味很好地融合在一起，留下一股咸味。

58%ALC./vol.

The First Impression	Breathe it in and feel the rush of spirit hit the senses and the nose prickle. Intense peat and smoke, maritime air, porridge oats, clean vanilla and toffee.
The Full-on Flavour	A burn of hot spirit and the mouth fills with smoke, feeling almost thick with oiliness before softening and becoming creamier, with nutty, briny and citrus notes.
FINISH	The mouth is almost in shock, numb and tingling. A sweetness mixes nicely with the slowly fading smoke, leaving a salty tang.

烟头雪利炸弹 /
Smokehead Sherry Bomb

48%ALC./vol.

第一印象　厚厚的泥煤味让你知道你手中有一瓶毫不含糊的艾雷岛单一麦芽威士忌。然后是一种水果味的、奶油感的、柔软丰富的橡木味，抚慰人心的奶油冻以及香草味。但先别觉得舒坦——这只是开始。

完全的风味　把自己绑起来等待油质黑色烟雾的袭击吧！它还带着浓重的覆盖口腔的泥煤感。随之而来的是一种甜而黏稠的户外烧烤香蕉味，（是的，真的）扑鼻的小葡萄干和干果味。告诉过你了，这是不一样的。

尾韵　泥煤味回来了，就像你曾希望或恐惧的那样。这一次，它带来了一些真正的西班牙雪利酒桶的橡木味，甜得令人欢喜令人忧。最后还有一个锐利的带着咸味的提示告诉你，你不是在跟寻常酒打交道。

IAN MACLEOD

DISTILLERS

48%ALC./vol.	
The First Impression	Thick peat lets you know you've got a serious Islay single malt on your hands. Then a fruitier, creamier feel—soft rich oak, soothing custard and vanilla. But don't get comfortable—this is only the beginning.
The Full-on Flavour	Tie yourself down and wait for the hit of oily black smoke, with a rich heavy mouthcoating peaty feel. Then along comes a sweet sticky flavour of barbecued banana, (yes, really) tangy sultanas and dried fruits. Told you it was different.
FINISH	The peat returns, just as you hoped/feared it would. This time it brings some authentic Spanish sherry oakiness with it, pleasingly and disturbingly sweet. And one last sharp salty reminder that you're not dealing with the ordinary here.

艾雷男孩 / Islay Boys

艾雷男孩 Islay Boys（照片由艾雷男孩提供 Photo by Islay Boys）

艾雷男孩是来自波特纳黑文的麦凯·史密斯和来自夏洛特港的唐纳德·麦肯齐。两个地方都是传统的小渔村，位于著名的威士忌岛艾雷岛的林斯半岛上，而这两个男孩都是与流淌在他们血液里的泥煤威士忌一起长大的！

他俩首先是多年的朋友，两人都喜欢优质的泥煤威士忌，或许还会配上一瓶很棒的啤酒：传统"半杯"威士忌（一小杯威士忌和一杯啤酒），至今在林斯的酒吧里也经常有人点呢！

他们意识到了交织在他们身上的凯尔特人和维京人的历史，尤其是麦凯的家族，与中世纪以艾雷岛为基地的诸岛领主紧密联系，并且他俩试图创造出能让他们祖先引以为豪的烈酒和啤酒。他们信仰着岛民的格言：威士忌首先并且最重要的，是"帮助解决困难，并增加日常生活的乐趣"。

除此之外，他们还带来了意义深远的专业经验，用到他们的威士忌和精酿啤酒中：麦凯有着重大的国际项目管理经验——从战略发展到执行，还有帮助酒商从苏格兰对外出口商品的经验，并曾在北美、欧洲和亚太地区的核心市场销售苏格兰精酿烈酒和啤酒。

The Islay Boys are Mackay Smith, from Portnahaven, and Donald MacKenzie from Port Charlotte. Both are small traditional fishing villages on the Rhinns peninsula of the famous whisky island of Islay, and both boys grew up with peated whisky flowing in their veins!

They are long-term friends above all, and both enjoy a good peated whisky, with perhaps a great beer alongside: the traditional "half and half" (a dram of whisky with an accompanying beer) of their forebears on Islay, and still a regular order at the Rhinns bars!

They are conscious of their intermingled Celtic and Viking history—Mackay's family, notably, was intertwined closely with the medieval Islay-based Lords of the Isles—and they try to create both spirits and beers that their forebears would be proud of. They adhere to the Isles men adage that a whisky is first and foremost made to "help with the difficulties, and to increase the joys, of daily life".

Beyond that, they bring significant professional experience to their whiskies and craft beers: Mackay has significant international experience managing projects from strategy development to implementation, as well as supporting alcohol companies to export from Scotland, and has worked in sales of Scottish craft spirits and beer in key markets in North America, Europe and Asia Pacific.

　　他如今在公司所有制的艾雷艾尔啤酒厂担任首席酿酒师，在这里，他用运营管理和全球供应链方面的专业知识，同时操持着艾雷艾尔啤酒和艾雷男孩的业务。另一方面，由布鲁赫拉迪的吉姆·迈克伊万培训了威士忌生产技术的唐纳德，在法国时就拥有了非常广阔的分配原酒的经验，在那里，他被视为"威士忌先生"。他带来了基于市场的威士忌品牌管理，并用他在管理经销商关系方面的巨大经验，给艾雷艾尔啤酒和艾雷男孩赋予了深入彻底的市场营销和公关背景。

　　麦凯和唐纳德都笃信要在他们最大的经济承受范围内推出以味道为最重要考量的烈酒和啤酒，同时不忘保有一定意义的西部岛屿幽默感！

诸岛领主

　　凯尔·弗拉诺斯是群岛的维京国王，他是一名航海勇士，在9世纪晚期统治着辽阔、崎岖、荒芜的苏格兰西部海岸。他的继任者是群岛的凯尔特领主，他们在主岛艾雷岛的菲拉根基地统治着他们的领地。

　　11世纪末，挪威的维京国王"光腿"马格努斯三世统治着苏格兰群岛。他接受了苏格兰高地的传统短裙，因此有了这个绰号！

He is now the head brewer at the company-owned Islay Ales brewery on Islay, where he manages operations for both Islay Ales and Islay Boys, using his expertise in operational management and global supply chain. Donald, on the other hand, who was trained in whisky production by Jim MacEwan at Bruichladdich, has very extensive experience of distributing spirits in France, where he is seen as "Mr Whisky", bringing in-market whisky brand management, along with an in-depth marketing and PR background to both Islay Ales and Islay Boys, with his tremendous experience in managing reseller relationships.

Both Mackay and Donald believe in proposing flavour-foremost craft spirits and beers, at the most affordable price possible, while bearing in mind a certain sense of Western Isles humour!

Lords of the Isles

Ketill Flatnose was a Viking King of the Isles, a seafaring warrior who ruled the vast, rugged and wild western seaboard of Scotland in the late 9th century. His successors were the Celtic Lords of the Isles, who ruled their domain from their base at Finlaggan on our home island, Islay.

Magnus III "Barelegs", Viking King of Norway, ruled the islands of Scotland in the late 11th century. He adopted the Highland tradition of the kilt, hence his nickname!

弗拉诺斯混合威士忌 /
Flatnose Blended Scotch Whisky

含15种威士忌，其中30%是麦芽威士忌。2种艾雷岛威士忌来自东海岸的艾雷岛酿酒厂，还有一些来自斯佩赛和高地产区。全部使用波本桶陈（首次/再次灌注），没有雪利桶陈。所有的单一麦芽威士忌都陈年5年或以上。还有在混合威士忌中很不寻常的工艺——非冷凝过滤。

43%ALC./vol.

闻香	柔和、光滑，有少许胡椒味。
味觉	刚开始活泼，然后香料味道逐渐发展成被碾碎的黑胡椒味，整体被丰富而均匀的背景支撑着。
余味	绵长而持久，以柔和的甜味结尾。

Islay Boys

15 whiskies, 30% malt. 2 Islay whiskies are from east coast Islay distilleries, with others being from Speyside and Highlands. All ex-bourbon (1st/refill) wood, no sherry. Single malts all 5 years or older. Very unusually for a blend—unchill-filtered.

43%ALC./vol.	
NOSE	Gentle and smooth with hints of pepper.
TASTE	Initial vivacity, then spices evolving towards cracked black pepper, the whole supported by a rich and even background.
FINISH	Long and persistant, ending in a gentle sweetness.

弗拉诺斯混合麦芽威士忌 /
Flatnose Blended Malt Scotch Whisky

含10种单一麦芽威士忌，瓶中1/3容量的威士忌是来自东海岸的艾雷岛的酒厂，其余由来自斯佩塞产区的8种单一麦芽威士忌构成，用于平衡烟熏味并带入一丝甜味。所用的单一麦芽威士忌都陈年5年以上。全部经过波本桶陈（首次/再次灌桶），没有雪利桶陈，非冷凝过滤。

43%ALC./vol.	
闻香	在蜂蜜味基底上的泥煤烟熏味扑面而来。
味觉	柔顺而持久的淡淡的泥煤味，与精致的甜调交相辉映，并辅以柑橘的果香。
余味	长而优美的烟熏味，并以温柔的泥煤甜调结尾。

Islay Boys

10 single malts, 1/3 bottle is from east coast Islay distilleries, with 8 others from Speyside to balance the smoke and bring a bit of sweetness. Single malts all 5 years or older. All ex-bourbon (1st/refill) wood, no sherry, unchill-filtered.

43%ALC./vol.	
NOSE	Initial peat smoke, over a honied background.
TASTE	Smooth yet persistant light peat, overlapping delicate sweet notes, which in turn are complemented by touches of citrus fruit.
FINISH	Long and delicately smoky, and ends in a gentle peated sweetness.

光腿艾雷岛单一麦芽威士忌/
Barelegs Islay Single Malt Scotch Whisky

小批量生产，来源的酒陈年6到8年不等，用再次灌注波本桶陈（通常为第二次灌注），非冷凝过滤。

46%ALC./vol.	
闻香	在余烬之上，是泥煤味的波浪，带着若有若无海风的气息。
味觉	新鲜直接的泥炭味，带有一丝甜味，与松脆的谷物味相平衡。
余味	悠长的结尾，持久的泥煤烟熏味，留下的口感很干净。

Islay Boys

Small batch, where age can vary from 6 to 8 years old. Refill bourbon cask maturation (generally second fill). Unchill-filtered.

46%ALC./vol.	
NOSE	Waves of peat, over embers, with sea breeze hints.
TASTE	Fresh direct peatiness, with a touch of sweetness, balanced by crisp cereal notes.
FINISH	Long finish, with persistant peat smoke, leaving a clean palate.

麦克达夫国际——艾雷之雾 /
MacDuff International—Islay Mist

　　艾雷之雾的独特之处在于它是源于艾雷岛本土的苏格兰混合威士忌。实际上，它是唯一得到苏格兰威士忌协会的允许，在名字中冠以"艾雷"一词的苏格兰混合威士忌——因为它的出处和历史：原本它就是在艾雷岛上，用拉弗格、格兰威特以及其他几种品牌的威士忌创造出来的，现在仍坚持着最初的理念和配方。

　　艾雷之雾原本是为玛加岱尔勋爵的儿子、继承人的21岁生日而创作，他的勋爵印章就装饰在瓶子上。这位继承人在1927年满21岁，当时的考虑是，从四面八方赶来艾雷屋，即今天艾雷岛上的玛加岱尔宅参加庆典的宾客们，会觉得本地威士忌烟熏味太重。这促成了独特的混合威士忌的诞生，人们就用本地拉弗格单一麦芽威士忌，加入大陆产的威士忌，创造出一种更丰满、更平衡以及更完整的特性，同时也有明显的艾雷风格的烟熏味——就像今天这样。

　　1987年，艾雷之雾与拉弗格一起被百龄坛的希拉姆·沃克收购，但到1992年，艾雷之雾在百龄坛的等级制度中落后了，并与两个历史悠久的格拉斯哥品牌一起被出售：劳德士和麦尼士。买家是苏格兰私人公司麦克达夫国际，该公司通过坚守威士忌的高标准，培育这个老品牌，使其恢复了昔日的辉煌。

Islay Mist is unique in the fact that it is a blended Scotch whisky originating from Islay. In fact, it is the only blended whisky, allowed by the Scotch Whisky Association, that is able to use the word "Islay" in its name—due to the provenance and history: Originally it was created on Islay using Laphroaig, Glenlivet and a few other whiskies, the original concept and recipe is still adhered to.

Islay Mist was created for the 21st birthday of the son, and heir of Lord Margadale, whose "Lord of the Isles" seal adorns the bottle. He turned 21 in 1927, and the considerations were, that the guests arriving from near and far for the celebration at Islay House, then Margadale residence on Islay, would find the local whiskies too smoky. This resulted in the creation of a unique blend using local Laphroaig single malt, but balancing with mainland whiskies creating a more rounded, balanced and fuller character, yet with the obvious Islay-smokiness—just like today.

Together with Laphroaig, Islay Mist was taken over by Hiram Walker (Ballantine's) in 1987, but by 1992 Islay Mist had fallen behind Ballantine's in their hierarchy and were sold off together with two historic Glasgow brands; Lauder's and Grand Macnish. The buyer was the private Scottish company MacDuff International who have since nurtured the old brand back to its former glory by guarding the high standards of the whisky in the bottle.

艾雷之雾原始 / Islay Mist Original

暗色的主标签完美地封装并体现了艾雷之雾的迷雾和烟熏感。足部的标签完美地描述了它："一种不会弄错的、柔和的艾雷岛烟熏味，还有甜麦芽、芳醇橡木味和艾雷岛海风的复杂融合。"

40%ALC./vol.	
闻香	泥煤烟熏味、大麦芽味以及耐嚼太妃糖味的微妙而诱人的平衡。
味觉	泥煤味是主调，但被甜麦芽和橡木味的奶油般的复杂融合体柔化了，带着少许海边的调调。
余味	整体被柔和的泥煤味糅合，被水果调的麦芽味和橡木味围绕。

The dark main label perfectly encapsulates and embodies the mist and smoke of Islay Mist. It is perfectly described on the foot label as "an unmistakable, yet gentle Islay-smokiness with a complex fusion of sweet malt, mellow oakiness and Islay sea breeze".

40%ALC./vol.	
NOSE	A subtle and seductive balance of smoky peat, barley malt and soft chewy toffee.
TASTE	The peat is dominant but mellowed with a creamy complex fusion of sweet malt and oak, with just a sprinkle of seaside.
FINISH	The full body is massaged by soft peat rounded by fruity maltiness and gentle oakiness.

失落蒸馏厂公司——洛西特 /
The Lost Distillery Company—Lossit

失落蒸馏厂公司是一家独立的精品苏格兰威士忌公司。他们痴迷于工艺，在威士忌的质量方面毫不妥协。他们的使命是创造传奇威士忌的现代诠释，而这些威士忌有着近一个世纪前的威士忌蒸馏工艺。他们用以下这段话来描述在做的事：

我们所做的事情没有什么神奇的公式。我们没有堆满被遗忘的旧威士忌的仓库，没有秘方或DNA分析，也没有计划重开这些失传的酿酒厂。我们所作所为的答案就在历史书中……

这个过程从我们格拉斯哥大学的迈克尔·莫斯教授领导的档案小组开始。关注着影响了这些长期失传的威士忌原始个性的10个关键因素，档案小组在证明原液最后一次蒸馏出来的味道方面发挥了关键作用。虽然我们可以说，生产过程中有许多因素对威士忌的最终口感至关重要，但我们只关注认为最重要的10个因素。根据研究中获得的信息，我们能够对这些威士忌曾经的大致轮廓做出确定假设。

The Lost Distillery Company is an independent boutique Scotch whisky company. They are obsessive about craft and uncompromising when it comes to whisky quality. Their mission is to create present day expressions of legendary whiskies that belonged to the craft of whisky distilling almost a century ago.

They describe what they do with these words:

There is no magic formula to what we do. We don't have a warehouse full of old forgotten whisky, we don't have a secret recipe or DNA analysis and we don't have plans to reopen any of these lost distilleries. The answer to what we do lies in the history books...

The process begins with our Archiving Team led by Professor Michael Moss from the University of Glasgow. Focusing on the 10 key components that influenced the original character of these long lost whiskies, the Archiving Team play[s] a pivotal role in evidencing how that spirit might have tasted when it was last distilled. While we could argue that there are a number of elements of the production process that are pivotal to the final taste of the whisky, we focus on what we consider to be the ten most important. Depending on the availability of information from our research, we are able to make certain assumptions as to what the profile of these whiskies might have been.

这10个关键因素是：

1. 年代——最后一次蒸馏的日期非常重要。与大多数制造业一样，时尚和生产流程也会发生变化。机械化提高了这一过程的一致性，而铁路的扩建则支持了更大酒厂的建设。

2. 酿造地点——邻近的酿酒厂可能使用了类似的水源、大麦和酵母。他们可能分享了专业技术，至今活跃于仍在进行生产的酿酒厂中。

3. 水——用于酿造原液的一个核心成分，也用来稀释产品的装瓶强度。水质是软是硬？矿物含量是多少？

4. 大麦——大麦最重要的方面是酚含量。这些大麦生长在哪里？是当地的吗？使用了哪些大麦品种？产量有多稳定？

5. 酵母——为何一些酵母面包比其他的好吃？为何一些面包师要保留面种几十年？酵母在生产过程中起了作用并且对最终成品产生影响。

6. 泥煤——大麦芽是否经过泥煤烘烤？用了多少泥煤？是本地的吗？这些如何转化为最终成品中的酚质含量？

7. 麦芽浆桶——它是用什么材料做的？它是开着的还是关着的？温度是怎么控制的？不稳定的温度会抑制酵母的活性。

The 10 Key Components:

1.Era—The date of the last distillation is critically important. As with most manufacturing businesses, fashions and processes change. Mechanisation brought increased consistency to the process, while expansion of the railways sponsored the construction of much bigger distilleries.

2.Locality—Neighbouring distilleries may have used similar sources of water, barley and yeast. They may have shared expertise that still survives today in working distilleries.

3.Water—A core ingredient used to make the spirit and also to dilute the product to bottling strength. Was the water soft or hard? What was the mineral content?

4.Barley—The most important aspect of the barley is the phenolic content. Where was the barley grown? Was it local? Which strains of barley were used? How consistent was the yield?

5.Yeast —Why is some sourdough bread better than others? Why do some bakers retain a starter dough for decades? Yeast matters in the process and ultimately has an impact on the final product.

6.Peat—Was the malted barley peated or unpeated? How much peat was used and was it sourced locally? How did this translate to the phenol content of the finished product?

7.Mash Tun—What material was it constructed from? Was it open or closed, and how was the temperature controlled? Volatile temperatures would inhibit yeast activity.

8.发酵槽——这些过去几乎无一例外用花旗松制作，选择它们是因为有直行纹理以及结节很少。虽然一些酒厂仍在使用这样的发酵槽，但大多数酒厂已改用不锈钢的发酵槽，不会赋予成品任何个性。

9.蒸馏器——蒸馏器的形状和大小深刻地影响了整个原液的个性。举个例子，一个相对较小的矮胖蒸馏器通常会让铜和原液之间有更多的接触，这意味着它会产生更重、更黏稠的原液。

10.木材——在生产出来后，哪种树木被用于储存或者运送这些威士忌到目的地？这影响到了最后的风味吗？在装威士忌之前，桶里装过什么？这对威士忌的口味会有显著影响。

研究过程的第三部分可能是最有趣也是最具挑战性的，我们称之为"辩论"。我们的档案小组成员和威士忌制造商，连同一组精选的"鼻子"，试图把摆在他们面前的证据活生生地呈现出来。他们把来自不同酿酒厂的单一麦芽威士忌混合在一起，加入了不同的口味，调整了成分，使之既符合档案员的证据，又符合威士忌制造商的解释。这一过程需要相当多的时间和经验，但只有当每个人都对结果感到满意时，这种威士忌才会得到失落蒸馏厂公司的认可——这时，我们为创造了失传已久的传奇威士忌的现代诠释而高兴。

8. Wash Back—These would have been made almost exclusively from Douglas Fir; chosen for its straight grain and lack of knots. While some distilleries still use these, most have converted to stainless steel versions that impart no character to the product.

9. Still—The shape and size of the still deeply influence the overall character of the spirit. For example, a smaller dumpy still will typically allow more contact between the copper and the spirit meaning that it produces a heavier, more viscous spirit.

10. Wood—After production, what type of wood was used to store or transport the whisky to its destination? Did this have an impact on the final flavour? What did the barrel have in it before it was used for whisky? This would have had a significant effect on the whisky's taste.

The third part of our process is arguably the most interesting and challenging. We call this "The Debate". Our Archivists and Whisky Makers, along with a panel of selected "noses", attempt to bring to life the evidence before them. They create a blend of single malts from different distilleries and with different flavour profiles, tweaking the composition to sit easily with both the evidence of the archivist and the interpretation of the whisky makers. This process takes considerable time and experience, but only when everyone is comfortable with the result does the whisky receive The Lost Distillery Company seal of approval—when we are happy that we have created a present day interpretation of that long lost whisky legend.

洛西特 / **Lossit**

创始人　马尔科姆·麦克尼尔（1785—1850），地主，农场主和酿酒师。

历史厂址　巴利格兰特，艾雷岛。

经营时间　1817—1867年，因被孤立、过时而关闭。

洛西特的图画 Lossit's drawing

Founder	Malcolm Macneill (1785—1850), Landowner, farmer & distiller.
Historical Location	Ballygrant, Islay.
Operated	1817—1867, closed due to being isolated and outdated.

洛西特经典 / Lossit Classic

43%ALC./vol.

闻香	泥煤、梨和杏仁牛奶的气息。
味觉	优雅的泥煤和胡椒味。
余味	烟熏味蜿蜒成一个长长的结尾。

The
LOST
DISTILLERY
Company

43%ALC./vol.	
NOSE	Peat, pear & almond milk.
TASTE	Elegant peat & pepper.
FINISH	Smoke flows to a long finish.

洛西特档案管理员 / Lossit Archivist

46%ALC./vol.

闻香	开始的泥煤烟熏、酿酒水果以及麦芽香——相当微妙，还带有淡淡的菠萝、香草和黄油烤面包的香气。
味觉	口感顺滑，奶油味扑鼻的开头，将人带到一堆冒烟的篝火旁，有一波又一波柔和的香料味道流淌在烟雾的气息中。
余味	姜饼、烤葡萄柚、绵长、烟雾缭绕、复杂——正当你认为它在消退时，香料的律动又回来了。

The
LOST
DISTILLERY
Company

	46%ALC./vol.
NOSE	From the beginning its peat smoke, wine fruit and malty notes—this rather subtle nose also has hints of pineapple, vanilla and buttered burnt toast.
TASTE	The palate has a smooth and creamy tangy start, leading to a smoky bonfire with waves of gentle spice flowing in between the puffs of smoke.
FINISH	Gingerbread, grilled grapefruit and its very long, smoky and complex—just when you think its fading, the waves of spice are back.

洛西特佳酿 / Lossit Vintage

46%ALC./vol.	
闻香	泥土般的海洋气息中夹杂着雪茄叶和阵阵烟熏味，黑巧克力和温暖夏夜的复杂气息。
味觉	香甜的烤梨在糖浆中炖着，接着是淡淡的烟熏和巴西坚果味，口感油腻，然后是香辛味。
余味	甜度达到漂亮的平衡，香料韵味在很长一段时间后才会褪去。

THE LOST DISTILLERY *Company*

46%ALC./vol.	
NOSE	Earthy maritime air with cigar leaves and a burst of smoke, dark chocolate and complex hints of a warm summer evening.
TASTE	Sweet baked pears stewing in syrup followed by subtle butts of smoke with brazil nuts, oily on the palate and then spicy notes come through.
FINISH	A beautiful balance of sweetness, spice only fades after a long time.

佳酿麦芽威士忌公司 /
Vintage Malt Whisky Company

　　佳酿麦芽威士忌公司是由布莱恩·克鲁克于1992年创立的，他在威士忌行业干了多年，在此之前是苏格兰一家著名酿酒厂的出口总监。他的目标是从自己祖国的最好的酿酒厂里，生产出一系列麦芽威士忌，并将其提供给世界各地的独立葡萄酒和烈酒进口商。

　　事实证明，布莱恩的方法非常受客户欢迎，27年来，该公司已向全球30多个国家出口了1000多万瓶麦芽威士忌，如今已被公认为苏格兰领先的独立装瓶商之一。佳酿麦芽威士忌公司现今由布莱恩的儿子安德鲁·克鲁克经营，在他的管理下，公司业绩获得了巨大的增长，同时他将父亲的愿景继续发扬光大。

　　苏格兰高地群岛威士忌公司旗下的品牌于1997年加入该集团。2005年，公司收购了艾雷风暴的生产商CS James & Sons。

　　该集团的获奖品牌包括菲拉根、艾雷客和艾雷风暴，它们都是来自苏格兰西海岸的艾雷岛的单一麦芽威士忌，而酷选大师则是一个极好的单桶瓶装系列。

THE
VINTAGE MALT WHISKY
COMPANY LIMITED

The Vintage Malt Whisky Company was founded in 1992 by Brian Crook, after many years in the whisky industry as Export Director for one of Scotland's best known distillers. His aim was to produce a range of malt whiskies from his country's finest distilleries and make them available to independent wine and spirit importers throughout the world.

Brian's approach proved so popular with customers that, 27 years on, the company has exported more than 10 million bottles of malt whisky to more than 30 countries across the world, and is now recognised as one of Scotland's leading independent bottlers. Vintage Malt is now run by his son, Andrew Crook, under his management the company has seen huge growth as he continues to take his father's vision forward.

The brands owned by Highlands & Islands Scotch Whisky Co. joined the group in 1997, and in 2005 the company acquired CS James & Sons, producers of Islay Storm.

The group's award winning brands include Finlaggan, the Ileach, and Islay Storm, all single malts from the island of Islay on the west coast of Scotland and the Cooper's Choice, a superb range of single cask bottlings.

艾雷风暴 / Islay Storm

40%ALC./vol.	
闻香	刺鼻的泥煤烟味，酚质气息，带着一点海风的气味。
味觉	甜泥煤味，一些碘酒和海盐的味道与香草和香料的味道达成平衡。
余味	一波又一波的烟熏味，非常悠长。

40%ALC./vol.	
NOSE	Pungent peat smoke. Phenolic, with a little coastal breeze.
TASTE	Sweet peatiness. Some iodine and sea salt balance with vanilla notes and spice.
FINISH	Waves of smoke, very long.

艾雷客 / The Ileach

40%ALC./vol.

闻香	年轻的泥煤、草本气息，橡木、青草、海洋气息。
味觉	极为柔顺的麦芽味、泥煤味、辛辣香料味、单宁酸和一丝水果味。
余味	微弱的碘酒味，海边的气息、麦芽味。

40%ALC./vol.	
NOSE	Youthful peat, herbal notes, oak, grassy, notes of the coast.
TASTE	Sublimely smooth malty notes, peat, peppery spices, tannins, a touch of fruit.
FINISH	Faint iodine. Coastal notes, malt.

艾雷客桶强 / The Ileach Cask Strength

58%ALC./vol.

闻香	泥煤、草本、橡木、青草和海边的气息。
味觉	柔顺的麦芽味、泥煤味、辛辣香料味、单宁酸和蜂蜜的甜蜜味道。
余味	泥煤味消逝得慢，混入微苦的橡木味和香料味。

58%ALC./vol.	
NOSE	Peat, herbal, oak, grassy and coastal.
TASTE	Smooth malty notes, peat, peppery spices, tannins, honey sweetness.
FINISH	Peat dissipates slowly, bitterish oaky influence and spices.

菲拉根老式珍藏 / Finlaggan Old Reserve

40%ALC./vol.

闻香	带土腥味的泥煤烟熏味以及咸咸的海风气息。
味觉	刺激的泥煤烟熏味，耐嚼的甜麦芽、胡椒、焦油以及少许碘酒味。
余味	悠长而温暖。泥煤篝火的烟熏灰烬味道。

40%ALC./vol.	
NOSE	Earthy smoky peat and salty ocean breeze.
TASTE	Pungent peat smoke, chewy sweet malt, pepper, tar and a touch of iodine.
FINISH	Long and warming. Smoky ashes of the peat fire.

菲拉根艾伦摩尔 / Finlaggan Eilean Mor

46%ALC./vol.	
闻香	甜泥煤烟熏味、药味及一点海盐气息。
味觉	酒体丰满,带有辛辣的泥煤烟熏味。焦油和碘酒的味道。轻盈的油感和甜麦芽味。
余味	烟熏的余烬,长而稍干。

46%ALC./vol.	
NOSE	Sweet peat smoke. Medicinal notes. Touch of sea salt.
TASTE	Full bodied with spicy peat smoke. Tar and iodine. Light oiliness and sweet malt.
FINISH	Smoking embers. Long and drying slightly.

菲拉根桶强 / Finlaggan Cask Strength

58%ALC./vol.

闻香	刺激性的泥煤烟熏味、烟熏培根带着一点旧皮革的气味。
味觉	浓郁的甜烟味、碘酒、柠檬皮的味道带着一层漂亮的覆盖口腔的油感。一波波的柏油泥煤味。
余味	辛辣的泥煤味，烟尘和灰烬，悠长而温暖。

58%ALC./vol.	
NOSE	Pungent peat smoke. Smoky bacon with a touch of old leather.
TASTE	Rich sweet smoke. Iodine, lemon zest with a beautiful mouth coating oiliness. Waves of tarry peat.
FINISH	Peppery peat. Soot and ash. Long and warming.

菲拉根波特桶 / Finlaggan Port Finished

46%ALC./vol.	
闻香	甜泥煤烟熏味，烟熏培根和夏季水果气息。
味觉	更加丰盈的甜蜜烟熏味，带一些覆盆子和草莓的味道。
余味	开始的泥煤篝火的灰烬味，接着是甜味。

46%ALC./vol.	
NOSE	Sweet peak smoke. Smoky bacon and summer fruit.
TASTE	More rich sweet smoke. Touch of raspberries and strawberries.
FINISH	Ashes of the peat fire. Late sweetness.

菲拉根雪利桶 / Finlaggan Sherry Finished

46%ALC./vol.

闻香	浓郁的泥煤烟熏味和大而成熟的水果气息。
味觉	厚而黏稠的泥煤味以及丰盈的水果味，烟熏和海盐、橘子和干果的味道。
余味	开始的泥煤篝火的灰烬味，接着是甜味，悠长而温暖，有更多的烟熏和水果味。

46%ALC./vol.	
NOSE	Rich peat smoke and big ripe fruits.
TASTE	Thick chewy peat and rich fruity notes. Smoke and sea salt. Oranges and dried fruits.
FINISH	Ashes of the peat fire. Late sweetness. Long and warming. More smoke and fruit.

资料来源 / Sources of Data

◎ **蒸馏厂以及公司**

阿贝

阿德纳霍

道格拉斯莱恩

波摩

布鲁赫拉迪

布纳哈本

卡尔里拉

A.D.拉特雷——艾雷酒桶

佳酿麦芽威士忌公司

艾雷男孩

麦克达夫国际——艾雷之雾

齐侯门

拉加维林

拉弗格

失落蒸馏厂公司

圣水蒸馏者

伊恩·麦克劳德蒸馏者

奥歌诗丹迪

◎ **参考网站**

www.islayinfo.com

www.visitscotland.com

www.scotland-info.co.uk

www.islay.com

www.wikitravel.org

◎ **Distilleries and Companies**

Ardbeg

Ardnahoe

Douglas Laing

Bowmore

Bruichladdich

Bunhabhain

Caol Ila

A.D.Rattray—Cask Islay

The Vintage Malt Whisky Company

Islay Boys

MacDuff International—Islay Mist

Kilchoman

Lagavulin

Laphroaig

Lost Distillery Company

Elixir Distillers

Ian Macleod Distillers

Angus Dundee

◎ **Reference Websites**

www.islayinfo.com

www.visitscotland.com

www.scotland-info.co.uk

www.islay.com

www.wikitravel.org

致谢 / Acknowledgement

所有信息、标志和瓶装威士忌图片都是由书中所提到的蒸馏厂和公司所提供。

照片的版权也归于上述来源。

我要感谢所有来自酒厂、公司和公关机构的善良的人，没有他们的帮助，这本书就不可能出版。

◎ 特别感谢

首先，我要感谢我的好友覃雨洋为翻译这本书所做的一切努力。

我还要特别感谢齐侯门蒸馏厂的创始人兼董事总经理安东尼·威尔斯，在我参观酒厂期间给予我的热情欢迎和所有帮助。

感谢布莱恩·伍兹，关键饮料公司、失落蒸馏厂的创始人，我曾愉快地参观了他的公司和威士忌酒室。

感谢来自伊丽宾馆的伊丽莎白和伊丽，让我在访问艾雷岛期间，在埃伦港度过了愉快的入住时光。

最后，我要感谢出版社为制作本书作出的所有努力。

All the info, logos and pictures of whisky bottles are provided by the distilleries and companies.

Photo credits as it's mentioned under them.

I wish to thank all beautiful people from distilleries, companies and PR agencies, without whose help, this book will not be possible to be made.

◎ Special thanks

First I wish to thank my very good friend Hecate (Qin Yuyang) for all efforts she did to translate this book.

Also my special thanks goes to Anthony Wills, founder and Managing Director of the Kilchoman Distillery, for warm welcome and all help during my visit to the distillery.

To Brian Woods, founder of the Crucial Drinks, the Lost Distillery, for a nice time I had visited his company and Whisky Room.

To Elizabeth and Eilidh from Eilidh's Guest House for a pleasant stay in Port Ellen during my visit to Islay.

Finally, I wish to thank the publisher for all the efforts to make this book.

词语注释 / Notes to Words

English	中文	注释	Explanation
SS Tuscania	SS图斯卡尼亚	一艘运兵舰的名字	Name of a troopship
HMS Otranto	HMS奥特朗托	一艘运兵舰的名字	Name of a troopship
HMS Kashmir	HMS克什米尔	一艘运兵舰的名字	Name of a troopship
ACE (Additional Cask Enhancement)	风味增强	一种熟成方式：威士忌从原来所在的橡木桶里转到装过其他酒的桶里面继续熟成。	ACE is a process of maturing whiskies: following proper maturation, in selected casks that previously held wines or spirits other than the original casks
PX (Pedro Ximénez)	PX、佩德罗·西门内	一种雪利酒	A kind of Sherry
Horlicks	好利克	一种麦芽制成的热饮	A kind of hot drink made from malts
Palo Cortado	帕罗科塔多	一种雪利酒	A kind of Sherry
Moscatel	莫斯卡托	一种雪利酒	A kind of Sherry

English	中文	注释	Explanation
Sauternes	苏特恩（又名苏玳）	一种葡萄酒	Sauternes wine
QA Cask	QA 桶	美国白橡木桶，拉丁语全称为Quercus Alba	America white oak cask, full name in Latin is Quercus Alba
Cask Strength	桶强	原桶装强度	The level of alcohol-by-volume (ABV) strength remain the same as in cask
bottle-aged	瓶龄	装瓶后熟成的	Aged in bottle
Sherbet Fountains	果子露喷泉	一种饮料	A kind of drink
hogshead	猪头桶	一种较小的酒桶（图片来自维基百科）	A kind of relatively small cask. (picture by Wikipedia)

Tun   Pipe, Butt 1/2 tun   Puncheon, Tertian 1/3 tun   Hogshead 1/4 tun   Tierce 1/6 tun   Barrel 1/8 tun   Rundlet 1/14 tun

威士忌品鉴笔记（模板）/
Whisky Tasting Notes （Template）

威士忌品牌名称（Brand）：　　　　　产区（Produce Origin in）：

蒸馏厂名称（Distillery）：　　　　　年份（Productive Year）：

酒精度数（ABV）：　　　　　　　　种类（Type）：

价格（Price）：　　　　　　　　　　购买/受赠时间（Time of buying/accepting as gift）：

何处购买（Buy from）：　　　　　　　　赠送者（Benefactor）：

酒标（Label）：　　　　　　　威士忌酒图片（Picture of this Whisky）：

威士忌品鉴（Tasting of this Whisky）——

色泽（Colour）：中度饱满，柔和的金色。

闻香（Nose）：果香和花香，香甜的麦芽、蜂蜜，纸板箱，饱经沧桑的木头，石楠花和一些薄荷的味道。

品味（Taste）：淡淡的蜂蜜味，顺滑巧克力和油酥点心中散发出温暖的酒精的芬芳。

余味（Finish）：温暖，黑醋栗，尾韵渐渐变淡。

注释（Comments）：酒体圆润，加冰块后，会散发出香甜的麦芽味搭配巧克力下午茶时饮用。

品尝日期（Date of tasting）：2020.04.20

等级（Rating）：★★★★☆

威士忌品牌名称（Brand）：　　　　　产区（Produce Origin in）：

蒸馏厂名称（Distillery）：　　　　　年份（Productive Year）：

酒精度数（ABV）：　　　　　　　　种类（Type）：

价格（Price）：　　　　　　　　　购买/赠予时间（Time of buying/accepting as gift）：

何处购买（Buy from）：　　　　　　赠送者（Benefactor）：

酒标（Label）：　　　　　　　威士忌酒图片（Picture of this Whisky）：

威士忌品鉴（Tasting of this Whisky）——

色泽（Colour）：

闻香（Nose）：

品味（Taste）：

余味（Finish）：

注释（Comments）：

品尝日期（Date of tasting）：

等级（Rating）：☆☆☆☆☆

威士忌品牌名称（Brand）：　　　　　　产区（Produce Origin in）：

蒸馏厂名称（Distillery）：　　　　　　年份（Productive Year）：

酒精度数（ABV）：　　　　　　　　　种类（Type）：

价格（Price）：　　　　　　　　　　购买/赠予时间（Time of buying/accepting as gift）：

何处购买（Buy from）：　　　　　　　赠送者（Benefactor）：

酒标（Label）：　　　　　　　　威士忌酒图片（Picture of this Whisky）：

威士忌品鉴（Tasting of this Whisky）——

色泽（Colour）：

闻香（Nose）：

品味（Taste）：

余味（Finish）：

注释（Comments）：

品尝日期（Date of tasting）：

等级（Rating）：☆☆☆☆☆

威士忌品牌名称（Brand）：　　　　　产区（Produce Origin in）：

蒸馏厂名称（Distillery）：　　　　　年份（Productive Year）：

酒精度数（ABV）：　　　　　　　　种类（Type）：

价格（Price）：　　　　　　　　　　购买/赠予时间（Time of buying/accepting as gift）：

何处购买（Buy from）：　　　　　　赠送者（Benefactor）：

酒标（Label）：　　　　　　　　　威士忌酒图片（Picture of this Whisky）：

威士忌品鉴（Tasting of this Whisky）——

色泽（Colour）：

闻香（Nose）：

品味（Taste）：

余味（Finish）：

注释（Comments）：

品尝日期（Date of tasting）：

等级（Rating）：☆☆☆☆☆

威士忌品牌名称（Brand）：　　　　　　　产区（Produce Origin in）：

蒸馏厂名称（Distillery）：　　　　　　　年份（Productive Year）：

酒精度数（ABV）：　　　　　　　　　　种类（Type）：

价格（Price）：　　　　　　　　　　　　购买/赠予时间（Time of buying/accepting as gift）：

何处购买（Buy from）：　　　　　　　　赠送者（Benefactor）：

酒标（Label）：　　　　　　　　　　　　威士忌酒图片（Picture of this Whisky）：

威士忌品鉴（Tasting of this Whisky）——

色泽（Colour）：

闻香（Nose）：

品味（Taste）：

余味（Finish）：

注释（Comments）：

品尝日期（Date of tasting）：

等级（Rating）：☆☆☆☆☆

威士忌品牌名称（Brand）：　　　　　产区（Produce Origin in）：

蒸馏厂名称（Distillery）：　　　　　年份（Productive Year）：

酒精度数（ABV）：　　　　　　　　种类（Type）：

价格（Price）：　　　　　　购买/赠予时间（Time of buying/accepting as gift）：

何处购买（Buy from）：　　　　　　赠送者（Benefactor）：

酒标（Label）：　　　　　　　　威士忌酒图片（Picture of this Whisky）：

威士忌品鉴（Tasting of this Whisky）——

色泽（Colour）：

闻香（Nose）：

品味（Taste）：

余味（Finish）：

注释（Comments）：

品尝日期（Date of tasting）：

等级（Rating）：☆☆☆☆☆

威士忌品牌名称（Brand）：　　　　　产区（Produce Origin in）：

蒸馏厂名称（Distillery）：　　　　　年份（Productive Year）：

酒精度数（ABV）：　　　　　　　　种类（Type）：

价格（Price）：　　　　　　　　　购买/赠予时间（Time of buying/accepting as gift）：

何处购买（Buy from）：　　　　　　赠送者（Benefactor）：

酒标（Label）：　　　　　　　　　　　　威士忌酒图片（Picture of this Whisky）：

威士忌品鉴（Tasting of this Whisky）——

色泽（Colour）：

闻香（Nose）：

品味（Taste）：

余味（Finish）：

注释（Comments）：

品尝日期（Date of tasting）：

等级（Rating）：☆☆☆☆☆

威士忌品牌名称（Brand）：　　　　　产区（Produce Origin in）：

蒸馏厂名称（Distillery）：　　　　　年份（Productive Year）：

酒精度数（ABV）：　　　　　　　　种类（Type）：

价格（Price）：　　　　　　　　　　购买/赠予时间（Time of buying/accepting as gift）：

何处购买（Buy from）：　　　　　　赠送者（Benefactor）：

酒标（Label）：　　　　　　　　　威士忌酒图片（Picture of this Whisky）：

威士忌品鉴（Tasting of this Whisky）——

色泽（Colour）：

闻香（Nose）：

品味（Taste）：

余味（Finish）：

注释（Comments）：

品尝日期（Date of tasting）：

等级（Rating）：☆☆☆☆☆

威士忌品牌名称（Brand）:　　　　　产区（Produce Origin in）:

蒸馏厂名称（Distillery）:　　　　　年份（Productive Year）:

酒精度数（ABV）:　　　　　　　　种类（Type）:

价格（Price）:　　　　　　　　　购买/赠予时间（Time of buying/accepting as gift）:

何处购买（Buy from）:　　　　　　赠送者（Benefactor）:

酒标（Label）:　　　　　　　　　　　　　威士忌酒图片（Picture of this Whisky）:

威士忌品鉴（Tasting of this Whisky）——

色泽（Colour）:

闻香（Nose）:

品味（Taste）:

余味（Finish）:

注释（Comments）:

品尝日期（Date of tasting）:

等级（Rating）: ☆☆☆☆☆

威士忌品牌名称（Brand）：   产区（Produce Origin in）：

蒸馏厂名称（Distillery）：   年份（Productive Year）：

酒精度数（ABV）：   种类（Type）：

价格（Price）：   购买/赠予时间（Time of buying/accepting as gift）：

何处购买（Buy from）：   赠送者（Benefactor）：

酒标（Label）：   威士忌酒图片（Picture of this Whisky）：

威士忌品鉴（Tasting of this Whisky）——

色泽（Colour）：

闻香（Nose）：

品味（Taste）：

余味（Finish）：

注释（Comments）：

品尝日期（Date of tasting）：

等级（Rating）：☆☆☆☆☆

威士忌品牌名称（Brand）：　　　　　　产区（Produce Origin in）：

蒸馏厂名称（Distillery）：　　　　　　年份（Productive Year）：

酒精度数（ABV）：　　　　　　　　　种类（Type）：

价格（Price）：　　　　　　　　　　购买/赠予时间（Time of buying/accepting as gift）：

何处购买（Buy from）：　　　　　　　赠送者（Benefactor）：

酒标（Label）：　　　　　　　　　　威士忌酒图片（Picture of this Whisky）：

威士忌品鉴（Tasting of this Whisky）——

色泽（Colour）：

闻香（Nose）：

品味（Taste）：

余味（Finish）：

注释（Comments）：

品尝日期（Date of tasting）：

等级（Rating）：☆☆☆☆☆

威士忌品牌名称（Brand）:　　　　产区（Produce Origin in）:

蒸馏厂名称（Distillery）:　　　　年份（Productive Year）:

酒精度数（ABV）:　　　　　　　种类（Type）:

价格（Price）:　　　　　　　　购买/赠予时间（Time of buying/accepting as gift）:

何处购买（Buy from）:　　　　　赠送者（Benefactor）:

酒标（Label）:　　　　　　　　威士忌酒图片（Picture of this Whisky）:

威士忌品鉴（Tasting of this Whisky）——

色泽（Colour）:

闻香（Nose）:

品味（Taste）:

余味（Finish）:

注释（Comments）:

品尝日期（Date of tasting）:

等级（Rating）: ☆☆☆☆☆

威士忌品牌名称（Brand）：　　　　产区（Produce Origin in）：

蒸馏厂名称（Distillery）：　　　　年份（Productive Year）：

酒精度数（ABV）：　　　　　　　种类（Type）：

价格（Price）：　　　　　　　　　购买 / 赠予时间（Time of buying/accepting as gift）：

何处购买（Buy from）：　　　　　　赠送者（Benefactor）：

酒标（Label）：　　　　　　　　威士忌酒图片（Picture of this Whisky）：

威士忌品鉴（Tasting of this Whisky）——

色泽（Colour）：

闻香（Nose）：

品味（Taste）：

余味（Finish）：

注释（Comments）：

品尝日期（Date of tasting）：

等级（Rating）：☆☆☆☆☆